Reflections on Roadkill between Mobility Studies and Animal Studies

Matthew Calarco

Reflections on Roadkill between Mobility Studies and Animal Studies

Altermobilities

palgrave
macmillan

Matthew Calarco
Department of Philosophy
California State University, Fullerton
Fullerton, CA, USA

ISBN 978-3-031-30577-1 ISBN 978-3-031-30578-8 (eBook)
https://doi.org/10.1007/978-3-031-30578-8

This Palgrave Macmillan imprint is published by the registered company Springer Nature Switzerland AG.
The registered company address is: Gewerbestrasse 11, 6330 Cham, Switzerland

Acknowledgments

I am indebted to two anonymous reviewers for the press who provided extraordinarily helpful feedback. I am also indebted to Amy Invernizzi for her sustained support of the project and expert handling of the manuscript. Finally, I wish to thank Brianne Donaldson for her faith in and encouragement of my work on this issue.

An earlier version of portions of Chap. 2 appeared as "Claimed by Roadkill," in *Feeling Animal Death: Being Host to Ghosts*, edited by Brianne Donaldson and Ashley King (New York: Rowman & Littlefield, 2019), 75–79. I am grateful to Rowman & Littlefield for permission to reproduce material from this chapter, all rights reserved.

Contents

CHAPTER 1

How (Not) to Look

Abstract Roadkill is a recurrent but often unthought feature of modern life. It is ubiquitous on urban, suburban, and rural roadways; it is a common theme on popular television shows and in horror movies; there are even cookbooks that extol the virtues and pleasures of eating roadkilled animals. Yet, consideration of the broader significance of the myriad ontological, social, ethical, and political issues related to roadkill has largely gone missing from mainstream scholarship and activism. This neglect persists even in fields such as mobility studies and animal studies that would otherwise seem to have a vested interest in the topic. Chapter 1 examines two systems of power that lie at the heart of the contemporary problem of roadkill: hyperautomobility and anthropocentrism. It is argued that examining these two systems in tandem is essential for overcoming the general neglect of roadkill in contemporary discourse and activism and for generating genuine alternatives to dominant systems of mobility.

Keywords Roadkill • Hyperautomobility • Anthropocentrism • The human • Mobility studies • Animal studies

Roadkill is a recurrent but often unthought feature of modern life. It is ubiquitous on urban, suburban, and rural roadways; it is a common theme on popular television shows and in horror movies; there are even

M. Calarco, *Reflections on Roadkill between Mobility Studies and Animal Studies*, https://doi.org/10.1007/978-3-031-30578-8_1

cookbooks that extol the virtues and pleasures of eating roadkilled animals. Yet, consideration of the broader significance of the myriad social, ethical, and political issues related to roadkill has largely gone missing from mainstream scholarship and activism. This neglect persists even in fields such as mobility studies and animal studies that would otherwise seem to have a vested interest in the topic. In this book, I argue that roadkill should be brought to the foreground of current discussions among scholars and activists in these fields and that the problems surrounding roadkill form a uniquely important site from which to understand and contest the machinations of the dominant social order. I argue further that a careful examination of the phenomenon of roadkill can help both to uncover the hidden violence of contemporary human-centered systems of mobility and to generate genuinely liberatory *altermobilities*—that is, alternative modes of mobility—that contest and reconstitute the extant social order in a less anthropocentric form.

Although this book contains a heavy dose of theory and is addressed primarily to an academic audience, it is at the same time a deeply personal and intimate project. In reflecting on this issue over the course of the past several years, I have come to recognize that encounters with roadkill have been a persistent and formative presence in my personal life since childhood.[1] Indeed, some of the earliest and most intense memories I can recall concern roadkill. Very early in my childhood, when I was perhaps five years old, I remember being inside my house and hearing terrible noises on the side of the house by the driveway. I quickly ran to the door to go outside and see what had happened. I then heard my father yelling to my mother to take me back in the house and not let me look outside. I went back in as I was told, but as I did so I glanced out the side window of the house toward the driveway. I saw my father kneeling down by the car and reaching for something near the front tire. As he stood up, I saw that he was holding a badly disfigured kitten that our family cat had recently delivered. As I later surmised, the kittens and their mother were temporarily using our car for shelter and shade, but the kittens were too young to know to get out of the way of the car when it was moving. The entire litter

[1] The word *reflecting* has increasingly imposed itself on me in the writing of this work. The Latin stem of this word, *flectere* (to bend, bend back, turn, curve, prevail on, soften), has proven itself to contain within it nearly everything I wish to say about roadkill and altermobilities. Although I will not try the reader's patience with wordplays and etymological musings throughout the text, I hope the multiple senses of this root will be borne in mind, along with related terms in which this root is present, such as deflect, inflect, genuflect, and so on.

had been run over when my father backed the car out of the driveway, and he was trying to remove them before I could see what had happened. Like me, my father adored those cats (and animals more generally). I could discern from his posture and heavy, deliberate movements that he was devastated. My family didn't talk afterward about what had happened. I had to piece together the details for myself.

Not long after that, I was driving with my father down the main thoroughfare in our town. Ahead in the road, I saw an animal moving in an unusual manner across the road. As we drove closer, I saw that it was a kitten (again). The kitten had had the back half of its body run over by a car and was trying desperately to drag itself off the road and out of traffic to the median. I didn't say anything as we got closer to the injured animal, but my father suspected I had seen the worst. He extended his arm toward me, gently covered my eyes with his hands, and said "Don't look, Matty boy." We drove on, in silence, neither one of us knowing what to say. My family didn't talk about this occurrence either, nor about the countless other roadkilled animals we encountered in the coming years.

There was considerable wisdom in my parents' gesture of trying to guard me while young from those terrible realities. As a child, I was very sensitive regarding animals, and they rightly discerned that being directly and prolongedly exposed to roadkill would have been too much for me to bear. I am confident that, were I to find myself in similar circumstances, I would make similar efforts to occlude such harsh sights from a child's gaze. Even for adults, the reality of roadkill is probably not something on which one ought to train one's attention in a sustained way for any significant period of time. For my part, I readily acknowledge that at times I have had to forcibly remove myself from looking at roadkilled animals and from examining the realities surrounding their deaths. The fact of the matter is that, for anyone who cares even minimally about animals, sustained consideration of roadkill can easily become psychologically overwhelming.

At some point, though, it becomes necessary to create the space to reflect on and respond to roadkill more directly than has been the case for so long now.[2] A justifiable response of avoiding too sustained and too

[2] With the phrase "for so long now," I am echoing the words of Jacques Derrida (2008), who speaks of his being tracked throughout his life by questions and issues concerning animals "*depuis le temps*," a phrase that carries a cluster of meanings relevant to the present project (see the translator's note on this phrase: Derrida 2008, 162, n. 2).

direct a confrontation with terrible realities ought not lapse into a complete avoidance of the issue at hand.[3] After all, to turn away from the reality of roadkill entirely and let our attention run over and past it is precisely what the established order encourages us to do with the animals who meet their death yearly by the millions on our roadways. We are taught to avoid truly seeing the roadkill that is strewn across our roads—or if we do happen to see it and are bothered by it, we have learned to dismiss it as an unfortunate consequence, as collateral damage, of a modern way of life that is ultimately non-negotiable and inevitable. This book is offered as but one effort among a growing number of others to resist that seeming inevitability. I focus on roadkill, then, not to be morbid but to articulate other possibilities for living (and dying), for moving (and staying put), in common with our animal and other more-than-human kin.

* * *

The main arguments presented in the chapters that follow took many years to come to fruition. For, even though I have been actively involved in alternative practices of mobility and animal activism since my teen years (I am nearing fifty as I write these words), the significance of roadkill largely escaped my critical attention and had little impact on my daily practices or activism for many years. Part of the reason for this lacuna is that I was unwittingly caught up in systems of power that I did not fully appreciate and for which I lacked the critical tools and counter-practices to extricate myself. In the remainder of this introduction, I briefly discuss two of these systems of power, systems I will refer to as *hyperautomobility* and *anthropocentrism*. I take up this analysis with some trepidation and considerable humility, as it has become common in the so-called race for theory (Christian 1987) to make grand claims about uncovering *the* dominant mode (or modes) of power in contemporary society, whether in the form of governmentality, exposition, attention, control, or some other modality.[4] My analysis here carries no such ambition; it is neither exhaustive nor

[3] For further examination of the complications involved in (avoiding) looking at roadkill, see MacKay (2019).

[4] See Koelle (2012) for a related take on this issue. Like Koelle, I avoid developing here an exhaustive theory of roadkill; nevertheless, I do wish to point out what I take to be important connections among certain systems of power that help partially to disclose the event of roadkill. For related considerations of the complexities involved in discussing modern systems of mobility, see Nail (2022).

does it seek to determine the originary locus of social power. Rather, I wish only to suggest that hyperautomobility and anthropocentrism are systems of power that play an important role in the formation of contemporary subjectivities at the individual and collective levels and that a better understanding of these systems helps, in part, to illuminate the significance of roadkill in relation to the established order.

First, then, a word about hyperautomobility. While the concepts of *mass motorization* and *automobility* are often used to name the transformations in transportation that marked the onset of the twentieth century, beginning in the mid-1970s these systems underwent important intensifications and modifications in advanced capitalist nations (Jones 2008; Urry 2004; Featherstone et al. 2005). From 1900 to the mid-1970s, individual car ownership went from being the privilege of a handful of individuals to becoming customary for the majority of the population. After this time, the percentage of individuals owning cars becomes relatively stabilized but the usage of cars starts to accelerate. From 1975 to the present, the number of trips that people take with their cars increases by over 50%, the length of those trips increases significantly, and the number of occupants per car on each trip is noticeably reduced. These changes in automobile use are tracked and explained with keen insight by sociologists Peter Freund and George Martin (2009). They argue that the changes in individual automobile use from 1975 to the present are so significant and distinct that they warrant referring to our age as one of *hyperautomobility*.[5] For Freund and Martin, hyperautomobility is best understood as "a deepening and broadening of personal car use to new levels" (478). As individual car use and the systems that support it grow rapidly from the mid-1970s onward, the automobile becomes increasingly central in the lives of individual drivers, and systems of automobile transport come to impact more individual lives. In short, in an era of hyperautomobility, "more people drive more miles in more complex vehicles" (481).

Freund and Martin explain that these intensifications of automobility are driven and supported by government policy, local forms of urban architecture, and a wide variety of industries involved in automobile

[5] My focus in this work is primarily on hyperautomobility, but I will make use of several related terms throughout the work, as hyperautomobility is not a stand-alone phenomenon but is an intensification of other patterns of mobility and emerges through and alongside a number of associated phenomena of the increased mobility characteristic of contemporary societies.

transportation; as such, systems of hyperautomobility and their supporting infrastructure can be remarkably resistant to change, even when they have obvious and widespread negative effects. Thus, even amid astonishing mortality rates, pervasive harmful health consequences, and catastrophic ecological effects caused by dominant systems of transportation, this dominant order is currently further expanding and intensifying with widespread public support and investment (Douglas et al. 2011; Lutz and Fernandez 2010).

The hard truth here is not just that systemic and infrastructural transformation is a difficult and lengthy process, with powerful socio-economic forces allied against it; that is, of course, quite true. The more vexing issue is that there is also very little desire on the part of most individual car owners and hypermobile individuals to institute and persist with the needed changes. We find ourselves in this paradoxical position because hyperautomobility is not something simply "out there," a system and set of structures that exist in a social world from which we as individuals are distinct. This system and set of power relations is now fully inside many of us; it governs us in manifold ways and has captured our subjectivity (Baerenholdt 2013). Billions of people now desire to drive their cars and strongly identify with this way of being-in and moving-through-the-world.[6] To transform these practices in a radical way would thus be tantamount to transforming many people's subjectivity in a radical way. For people who have little familiarity with other systems of mobility and other visions for being-in-the-world, such prospects of fundamental change pose a profound existential challenge.

Born and raised during the era delimited by Freund and Martin, I too was swept up in and captured by systems of hyperautomobility. Daily life with my family in our hometown of Escondido, California saw constant and extensive automobile use. I was shuttled around to school and social events, as little alternative transportation infrastructure was available and urban sprawl made it necessary to travel ever-longer distances. Like many American teenagers of my generation, I longed to have a driver's license

[6] See Meyer (2015), Sheller (2004), and Hui (2014) on this point. Peter Sloterdijk goes so far as to suggest that the automobile is "the sanctum of modernity; it is the cultish center of a kinetic world religion; it is the sacrament on wheels that lets us take part in something that moves faster than we do" (Sloterdijk 2020, 9). Sloterdijk is clearly cognizant of the fact that automobility of this sort is premised on a desired break by human beings from animal existence and that the car "makes the total world of highways into a home for us and makes us aware that we are called to something more than a half-animal pedestrian life" (9).

and own my own car. I began working at a young age to help pay for my car loan, gas, and maintenance; and once I turned sixteen years old and could drive independently, I drove myself everywhere and (because of low gas prices) thought nothing of regularly driving thirty-to-forty miles round trip to visit friends across town or go to the beach. Using a car for most of my daily tasks and activities became an entirely normal and nearly invisible part of my life.

It was not until I was exposed to the practices, politics, and theory of alternative mobilities that I started to grapple with the depth of my capture by the system of hyperautomobility. The "new mobilities paradigm" (Sheller and Urry 2006; cf. Randell 2020), which took form just around the time I began serious intellectual study, emerged in no small part in response to the system of hyperautomobility as well as to a rapid increase in mobility across a number of registers. Housed primarily in sociology but branching out into a number of related fields of discourse and movements for social change, the field of mobility studies has opened up a number of avenues for critical reflection and activism on contemporary modes of transport and movement, including various ontologies of mobility (Cresswell 2006; Sloterdijk 1987); the history of the automobile and other technologies of modern mobility (Kay 1997); environmental histories of mobility (Wells 2012); and the uneven social consequences and injustices of contemporary modes of mobility.[7] What this substantial body of research and activism has demonstrated is that the widespread intensification of mobility characteristic of our collective present is not just one phenomenon among others but one of the defining features of the age, one that affects and transforms everything from individual subjectivity and collective life to urban infrastructure and economic arrangements.

Despite the undeniable importance and transformative potential of this body of work, what remains missing from it is sustained analysis of how contemporary mobility impacts animal life, both in the form of roadkill and in related kinds of harm and death.[8] Such neglect of the effects of power on animals and the more-than-human world is an instance of the

[7]The issue of mobility justice will be examined at more length in Chap. 5, and I will engage with this literature there.

[8]There are important exceptions to this general trend, of course. What should be emphasized here, though, is that these exceptions primarily take the form of individual authors; the problem and theme of roadkill itself is largely ignored inside mobility studies (Swart 2015) and is treated in only a marginal way even in animal studies (a point I will discuss in what follows).

more general phenomenon of what I will refer to as *anthropocentrism*, the second system of power I wish to highlight in this introduction. The more common term used in animal studies for the overriding of the interests and lives of animals is *speciesism*, and I should note why I generally avoid employing this term before explicating the fuller meaning of anthropocentrism.[9] Coined by Richard Ryder (1989) and popularized by Peter Singer (2002) and related figures, speciesism is typically defined as an intellectual and illogical prejudice that denies moral consideration to animals on the premise that they do not belong to the human species; stated in reverse, the speciesist believes that all and only members of the biological species *Homo sapiens* deserve full moral standing and consideration. What makes speciesism a prejudice, according to its critics, is its dogmatic character and its inability to withstand critical intellectual scrutiny. Pro-animal critics of speciesism charge that the species boundary in itself is not a relevant or defensible moral marker and that the traditional markers of full moral standing found in human beings (such as sentience or subjectivity) are found in many animal species as well. Consequently, if the speciesist is to be logically consistent and avoid prejudice, then (the pro-animal theorist argues) moral standing must be extended to all beings—whether human or animal—who bear the relevant moral traits. Failure to enact such an extension is tantamount to sheer prejudice, on a par with other dogmatic "ism"s such as racism and sexism, which (on this account) are ultimately similar instances of overlooking the interests of people of color and women without sufficient justification.

Although, as I have just noted, the use of the concept of speciesism is common in the animal studies and animal ethics literature, I want to argue here that it fails adequately to explain why many animals are often granted little or no moral standing by the dominant social order; further, I believe it fails accurately to describe the situation of many marginalized human beings. A more helpful concept for understanding these power relations is that of anthropocentrism, which I take to refer to a set of ideas, structures, and practices that establish and reproduce the privileged status of those who are deemed to be fully and quintessentially human. Contra the

[9] I have discussed the distinction between speciesism and anthropocentrism in previous works (most recently, Calarco 2020). I do not want to assume, however, that readers will be familiar with my other writings (especially readers who come from outside animal studies and philosophy), so I provide a condensed overview of these concepts in the next four paragraphs. Readers familiar with this distinction can pass over these remarks.

underlying assumptions of the discourse surrounding speciesism, it is crucial to note that in Western cultural and intellectual traditions in particular, *the human* (from the Greek *ho anthrōpos*, from which is derived the term anthropocentrism) has only rarely and recently been taken to denote human biological species membership; more commonly—both historically and in the present—the concept of the human is understood to comprise only a select group of privileged individuals within that set. The human most commonly functions, then, as a *normative* ideal—that is, as a subject position that must be acceded to and enacted—rather than as a *biological* marker. The scope and extent of the human is maintained by excluding and denigrating certain "animal" and "nonhuman" traits and behaviors considered to be both internal and external to humanness. What is included and excluded under the rubric of the human shifts over time, of course, and group belonging expands and contracts depending on a number of factors; but one theme that remains deeply consistent throughout the history of this concept in the West is the general exclusion of animals and animality from humanity proper.

Anthropocentrism, then, is grounded on the notion that the human is exceptional in relation to animals and all other nonhuman beings. Characterized by purportedly unique and superior capacities and powers, the human justifiably becomes (according to anthropocentric logic) the focus and center of attention (hence the *centrism* of anthropocentrism). Anthropocentrism thereby functions as a kind of human narcissism, an attempt to grant greater importance, standing, and meaning to the human (and, to emphasize once more, *not* the biological species *in toto*) within nature and the cosmos as a whole. That there are certain points of continuity between human beings and animals has never been denied by the dominant intellectual and cultural sources of anthropocentrism. But what allows human exceptionalism and narcissism to be established and maintained is the claim to a human distinctive (or what is sometimes called an *anthropological difference*). Thus, even though humans share certain behaviors and characteristics with animals, anthropocentric logic posits that there must exist a sharp ontological rupture along certain axes wherein the human in its unique form emerges. Often, this (purported) rupture is marked by such capacities as mind or language, or sometimes to a cluster of emergent traits, that grant human beings their (again, purportedly) unique mode of existence.

This kind of exceptionalist ontology forms the deep background against which dominant normative, legal, and economic systems are structured. With the uniqueness of the human taken for granted, the circle of consideration is drawn tightly in line with the contours of the human and its interests. Full consideration need not be (and, in practice, rarely is) given to all members of *Homo sapiens* within an anthropocentric context, as not all human beings necessarily carry the unique traits of the human. Likewise, granting full or partial consideration to beings or species that are biologically nonhuman does little to disrupt the basic logic of anthropocentrism if that extension is based on such entities being similar to, or being in the best interests of, paradigm members of the class of the human. At an institutional level, the lack of standing for nonhuman beings leads in our age to most animals and more-than-human entities being reduced to property under the law and to commodities within the economy. Laws are designed with the interests of the human in mind, and markets are created and maintained to further the economic interests of the human alone.

It is this anthropocentric system of power—which includes a number of interrelated structural, material, infrastructural, and ideological components—that lies at the heart of the violence and injustice directed against animals, marginalized human beings, and others who are understood to fall outside the orbit of the subject position of the human. These power relations, then, are not mere prejudices or irrational beliefs (as the notion of speciesism would have it) but are instead deeply rooted ideological and material realities that form and shape individuals from the ground up. We should thus expect that countering this system of power will entail more than employing critical thinking skills or changing our beliefs. Rather, as I will argue throughout these chapters, anthropocentrism (much like dominant systems of mobility) can only be contested through a radical conversion in individual and collective ways of life. To be sure, such conversions involve changes in beliefs and the adoption of different modes of thought, but intellectual changes are but one part of a larger process of transformation.

Throughout my childhood years, my daily life was entirely and uncritically immersed within this set of anthropocentric structures. I ate meat, used animal byproducts, and participated unreflectively in countless social and economic activities that were predicated on the displacement, violation, and death of animals. Although I was extraordinarily fond of animals during these years, I had little understanding of how my way of life was destroying innumerable individual animals as well as the biophysical systems on which they depend for sustenance and well-being. Just before I became a teenager, my parents decided to adopt a vegetarian diet for health reasons. As they began to remove meat from our family meals, I recall insisting that I still wanted to eat "my hamburgers" (one of my favorite meals at the time). I was already, at that tender age, a devoted meat eater and had at least a passing awareness that giving up meat meant losing an important part of myself. My parents were understanding and didn't force me to eat strictly vegetarian food; shortly after this time, though, I voluntarily joined them in that way of eating. I had begun reading the vegetarian diet books my parents kept around the house, some of which contained descriptions of contemporary factory farming practices, replete with pictures and details of the litany of horrors to which farm animals are subjected. I became a committed vegetarian from that point forward based on both ethical and political reasons, although I had only the vaguest sense of what that commitment would entail.

Being a vegetarian set me on the path toward a series of changes in my everyday life. I began to rethink not just the food I ate but the products I used on a day-to-day basis, many of which I came to learn either incorporated animal byproducts or caused harm to animals through such means as testing or destroying their habitats. Over time, I came to understand that the egg and dairy industries treated animals in a cruel and entirely instrumental manner, which in turn led to me becoming a vegan in my teenage years. I then became more deeply involved in pro-animal discourse and activism, which broadened my understanding of anthropocentrism and helped me to appreciate just how deeply anti-animal violence permeates our daily lives and social institutions.

The problem of roadkill, however, remained in the background of the pro-animal discourse and activism in which I was involved. There was virtually no mention of roadkill or the effects of mobility on animals in any of the major works in the field, and there were few campaigns among the national or grassroots animal rights organizations with which I was involved to reduce roadkill or change driving habits. The problem of roadkill remained in the background of my personal life as well. Although I would experience profound feelings of grief when I occasionally struck animals while driving or in encountering roadkilled animals while out walking or cycling, I was curiously unable to join together my critical sensibilities about mobility and animals. Occasionally, some connection between the two would flash into my conscious awareness, but I was hesitant to look at the matter too carefully.

* * *

At some point, though, I began to look. I began to look, not in the sense of simply staring at roadkilled corpses (although some of that practice was involved) but more in the sense of trying to grasp the broader significance of roadkilled animals—their lives and deaths, their interrupted relations, and the deeper structures that made such deaths both so common and so insignificant for the dominant culture. As I will explain in the chapters that follow, it was only by bringing together alternative practices and critical reflections on mobility and animals that something of the being and meaning of roadkill began to emerge for me. What I have concluded and what I hope to demonstrate to the reader is that careful consideration of roadkill is a particularly helpful and productive way to bring mobility studies and animal studies into dialogue, both in order better to comprehend the event of roadkill itself as well as to recast some of the central stakes of both fields. In this vein, I argue that many of the currently popular strategies for achieving more just and equitable modes of mobility tend, from the joint perspective developed here, to reiterate uncritically a logic of sacrifice that allows for many animals to be framed as mere collateral of human-centered mobility systems. Conversely, by taking into account issues surrounding mobility and roadkill, I argue that animal studies is challenged to think about justice in ways that go beyond its traditional focus on the abolition of meat-eating, invasive experimentation, and other such forms of direct violence against animals and toward more complex questions having to do with indirect (infra)structural harms and injustices.

I should note also that, although the focus of this book will be primarily on what is colloquially called "roadkill," this term is far from unproblematic and does not capture the full scope of what is at stake in the discussion. As I argue in the next chapter, what we refer to as roadkill tends to confound standard ontological discourse—which is to say, it is hard to discern precisely what roadkill *is* or its social and ontological significance. At the same time, roadkill—understood in the common sense as an animal that is struck and killed by a moving automobile—is but one aspect of what is at issue in the present work. In view of the fact that a hypermobile culture makes use of numerous forms of transportation besides automobiles (such as planes, trains, ships, and so on) that also cause animals serious death and injury, it will be important to bear in mind other human-animal intersections and points of collision and contact beyond car-based roadkill.[10] Further, given the concept of anthropocentrism sketched above, we might anticipate that dominant systems of mobility disproportionately harm not just animals and other nonhuman beings but also dehumanized and subhumanized beings of all sorts. Thus, even if paradigm members of the class of the human are also harmed or killed by these systems, their lives are understood to matter and their deaths to be worthy of mourning. By contrast, the lives of large swaths of humanity are positioned by these systems to be sacrificeable, their deaths lacking in any kind of comparable significance. My aim, then, in the pages that follow is to attend to the ways in which anthropocentric systems of mobility produce these complex positionings of various marginalized beings, both human and more-than-human.

References

Baerenholdt, Jørgen O. 2013. Governmobility: The Powers of Mobility. *Mobilities* 8 (1): 20–34.

Calarco, Matthew. 2020. *Beyond the Anthropological Difference*. New York: Cambridge University Press.

Christian, Barbara. 1987. The Race for Theory. *Cultural Critique* 6 (1): 51–63.

Cresswell, Tim. 2006. *On the Move: Mobility in the Modern Western World*. New York: Routledge.

Derrida, Jacques. 2008. *The Animal That Therefore I Am*. Translated by David Wills. New York: Fordham University Press.

[10] Historian Gary Kroll proposes the term "snarge" (a word initially coined to refer to the remains of a bird that has been struck by a plane) be employed to refer to the wide range of collisions that occur between various kinds of "fossil-fuel-based human mobility and solar-based animal mobility" (Kroll 2018, 82). Although I will largely retain the term roadkill throughout this book, I share Kroll's concern to mark the variety of encounters, collisions, and points of contact between mobility systems and human and more-than-human lives/deaths. For further analysis of shipping-related collisions (especially in regard to whales), see Laist et al. (2001); on aircraft-animal collisions, see DeVault et al. (2013).

DeVault, Travis L., Bradley F. Blackwell, and Jerrold L. Belant. 2013. *Wildlife in Airport Environments: Preventing Animal-Aircraft Collisions Through Science-Based Management*. Baltimore, MD: Johns Hopkins University Press.
Douglas, Margaret J., Stephen J. Watkins, Dermot R. Gorman, et al. 2011. Are Cars the New Tobacco? *Journal of Public Health* 33 (2): 160–169.
Featherstone, Mike, Nigel Thrift, and John Urry, eds. 2005. *Automobilities*. London: SAGE.
Freund, Peter, and George Martin. 2009. The Social and Material Culture of Hyperautomobility: 'Hyperauto'. *Bulletin of Science, Technology & Society* 29 (6): 476–482.
Hui, Allison. 2014. Enthusiasm. In *The Routledge Handbook of Mobilities*, ed. Peter Adey, Mimi Sheller, David Bissell, Kevin Hannam, and Peter Merriman, 172–182. New York: Routledge.
Jones, David W. 2008. *Mass Motorization + Mass Transit: An American History and Policy Analysis*. Bloomington, IN: Indiana University Press.
Kay, Jane. 1997. *Asphalt Nation: How the Automobile Took Over America, and How We Can Take It Back*. New York: Crown Publishers.
Koelle, Alexandra. 2012. Intimate Bureaucracies: Roadkill, Policy, and Fieldwork on the Shoulder. *Hypatia* 27 (3): 651–669.
Kroll, Gary. 2018. Snarge. In *Future Remains: A Cabinet of Curiosities for the Anthropocene*, ed. Gregg Mitman, Marco Armiero, and Robert Emmett, 81–88. Chicago: University of Chicago Press.
Laist, David W., Amy R. Knowlton, James G. Mead, et al. 2001. Collisions between Ships and Whales. *Marine Mammal Science* 17 (1): 35–75.
Lutz, Catherine, and Anne Lutz Fernandez. 2010. *Carjacked: The Culture of the Automobile and Its Effect on Our Lives*. New York: Palgrave Macmillan.
MacKay, Carley. 2019. Through the Shadows of Roadkill. *Humanimalia* 11 (1): 128–140.
Meyer, John M. 2015. *Engaging the Everyday: Environmental Social Criticism and the Resonance Dilemma*. Cambridge, MA: MIT Press.
Nail, Thomas. 2022. What Is the Philosophy of Movement? *Mobility Humanities* 1 (1): 6–22.
Randell, Richard. 2020. No Paradigm to Mobilize: The New Mobilities Paradigm Is Not a Paradigm. *Applied Mobilities* 5 (2): 206–223.
Ryder, Richard. 1989. *Animal Revolution: Changing Attitudes Towards Speciesism*. Oxford: Basil Blackwell.
Sheller, Mimi. 2004. Automotive Emotions: Feeling the Car. *Theory, Culture & Society* 21 (4–5): 221–242.
Sheller, Mimi, and John Urry. 2006. The New Mobilities Paradigm. *Environment and Planning A: Economy and Space* 38 (2): 207–226.
Singer, Peter. 2002. *Animal Liberation*. New York: Ecco.

Sloterdijk, Peter. 1987. *Kopernikanische Mobilmachung und Ptolemäische Abrüstung*. Frankfurt: Suhrkamp.

———. 2020. *Infinite Mobilization: Towards a Critique of Political Kinetics*. Translated by Sandra Berjan. Medford, MA: Polity.

Swart, Sandra. 2015. Reviving Roadkill? Animals in the New Mobilities Studies. *Transfers* 5 (2): 81–101.

Urry, John. 2004. The 'System' of Automobility. *Theory, Culture & Society* 21 (4–5): 25–39.

Wells, Christopher W. 2012. *Car Country: An Environmental History*. Seattle: University of Washington Press.

CHAPTER 2

What *Is* Roadkill?

Abstract This chapter considers the ontological status of roadkill, which is to say, its basic nature and significance. It is suggested that traditional substance ontologies, as well as more contemporary relational ontologies and ethical frameworks, are inadequate for disclosing the nature of roadkill. In order to address this limit, Calvin Warren's analysis of ontological terror is employed to analyze the lives and deaths of beings who are understood to fall outside the orbit of the human. The chapter concludes with the idea that thinking and attending to roadkill as inhabiting such an outside requires relating to it as an event.

Keywords Substance • Face • Image • Outside • Terror • Event

I am riding my bicycle on a scorching hot day in Orange, California, at the edge of Santiago Canyon. Just ahead of me, I catch sight of a dead animal in the middle of the road. Checking for traffic and seeing none in the immediate vicinity, I bring my bicycle to a stop on the inside lane and briefly inspect the body. A rabbit. I lean down and try to remove the body from the sweltering blacktop. The body has been pressed into the road repeatedly by car tires, and the heat makes the flesh and fur stick to the pavement. As I pry the body loose, a leg becomes detached. I decide I

M. Calarco, *Reflections on Roadkill between Mobility Studies and Animal Studies*, https://doi.org/10.1007/978-3-031-30578-8_2

have to leave the leg behind for fear of being struck by cars approaching off in the distance. While holding the rabbit with one hand, I clumsily remount my bike and try to maneuver as quickly as possible over to the side of the road and back to the safety of the bike lane.

I then realize there is no grass on the side of the road (the usual resting place I seek for roadkill); there is only rock-strewn dirt, bordered by a fence. On the other side of the ivy-covered chain link fence, however, I spy lush green lawns. I lift the rabbit over the top of the fence and drop it as gently as I can on the other side. As the rabbit lands, a wave of grief comes over me. I say softly to the dead body on the other side of the fence: "I am so sorry." I repeat this utterance two more times, remount my bicycle, and begin slowly pedaling down the road.

It doesn't escape my attention that the rabbit I just encountered met its death mere yards from the Holy Sepulcher Cemetery, that it made its life there, amid the buried bodies and coffins of numerous dead human beings. By removing the rabbit from the road and placing it in the grass, by attending to its body and giving it what I hope is a more respectful resting site and a better end, am I ultimately hoping for the rabbit's resurrection? Even if it is only the resurrection of re-entering food chains and rejoining the general economy of life?

My mind drifts over to the cars and homes around me. Has anyone seen me? What must people think of a lycra-clad cyclist stopping in the middle of the road, picking up dead animals, placing them in the grass, and whispering to them? I have had occasion to ask myself this question more than once, for I have been engaged in this ritualized practice of roadkill removal for many years now. People undoubtedly see me and wonder what I am doing. Sometimes, I mentally rehearse answers to potential questions from passersby, hoping that I can muster a reasonable response that will reassure the interlocutor of my sanity. I have developed several standard answers over the years. But today when I reflect on what I am doing, no answer is forthcoming—the grief fails to lift, and no narrative I can muster offers insight or respite. Who was that rabbit? Who were its relations? What lives and worlds have been shattered here on this road? And what is at stake in these ritualized interactions with roadkill? Furthermore, what *is* roadkill? And what, exactly, am I doing with it? More important, what is it doing with me?

Substance

"What *is* roadkill?" is an infrequently posed question. In raising this question here, we are not seeking an answer in the form of a dictionary definition.[1] As noted in Chap. 1, such definitions can offer us an initial orientation into our problematic but will ultimately be insufficient for the task at hand. What is at issue in this chapter is trying to discern the *ontological* status of roadkill. What, exactly, *is* it? Does roadkill "exist" in any meaningful sense at all?

There is a venerable tradition in Western philosophy that traces the ontological status of a given entity back to its *power*. In Plato's *Sophist*, for instance, we find the Eleatic Stranger making the claim that "those entities which are" amount to nothing other than "δύναμις," or power (τὰ ὄντα ὡς ἔστιν οὐκ ἄλλο τι πλὴν δύναμις) (Plato 1997, 247e). Here δύναμις marks the capacity of an entity to make some kind of impact, to affect another entity or to be affected by it, to produce (ποιεῖν) some change in another entity or to suffer or undergo (παθεῖν) change at the hands of another entity. Were an entity to lack this δύναμις, lack any relation to ποιεῖν or παθεῖν, it could not, strictly speaking, be said to exist at all on this account; as such, it would not belong to the class of entities (τὰ ὄντα) that are. Quintessential examples of such beings with δύναμις include Aristotle's primary substances (πρῶται οὐσίαι) in regard to which varied characteristics are predicated but which remain ontologically and numerically the same despite certain variations (Aristotle 1987, *Categories*, 1b11 and 2a35). Primary substances are the kinds of entities that can both produce and undergo changes without being swallowed up or annihilated by them. Something of their singular δύναμις remains, despite the myriad relations they constitute and in which they participate.

Variants of substance ontologies have dominated Western ontology for millennia, but they are ill-suited for explaining the being of roadkill. Roadkill seems to hover on the edges of the world and to occupy a liminal, spectral site, somewhere between nothingness and full-bodied existence; it appears to have *some* force, but not the force or persistence traditionally associated with substances. To be fair, it is not just substance ontologies that fail to do justice to the being of roadkill; many of our contemporary critical and relational ontologies fare just as poorly. Can roadkill be said to

[1] See Desmond (2016, 143–144) for a general overview of roadkill from this more definitional perspective.

be shot through with "*différance*," structured by mediation and responsive relations with alterity? Can it properly be called a withdrawn "object" when it is so easily interiorized by a scavenger or flattened into the pavement to the point of unrecognizability? Is roadkill a "companion species" if it simply sits there unattended and disengaged and melts into the asphalt under the hot sun? Can a month-dead carcass decaying in the gravel on the roadside be rightfully called an "actor" in a network? Is roadkill "vibrant," delightful in its agential capacity? Here it will not do simply to take leave of these more vitalist and affirmative frameworks and align roadkill with so-called dark ontologies that stress decay, entropy, and annihilation, since roadkill circulates in an ontological realm that is more complex than these latter frameworks would allow us to glimpse. For roadkill, despite its lack of vitality (at least as this term is typically deployed in contemporary ontological discussions), seems nevertheless to carry *some* kind of force, power, or capacity that confounds the logic of δύναμις but is not simply reducible to cosmic, planetary, or ecological narratives of decay.[2]

Face

Perhaps roadkill would be better characterized using proto-ethical or pre-ontological terms rather than ontological categories. Perhaps, to borrow Emmanuel Levinas' terminology, roadkill has a "face" that would account for its uncanny ontological force. But a face, on Levinas' account, must be able to "*attend* its expression" (Levinas 1974, 55; see also Levinas 1969, 178). Are dead animals able to express themselves in the Levinasian sense? Do dead animals have faces? We are inclined to think of a face in the ethical sense as a brute fact, as something entirely mind-independent, as having the inherent expressive power to interrupt one's everyday mode of existence. Given this mind-independence, it would seem there are correlative necessary and sufficient conditions that define a face and explain its force. Thus, if roadkill lacks the force to interrupt the projects of the vast majority of people (a fact that would violate both the concept's necessary and sufficient conditions), this would seem to indicate that dead animals simply do not have faces. Phenomenologically speaking, we might say that roadkill just does not have the power to interrupt standard modes of intentionality and subjective enjoyment.

[2] On this point, see also Lulka (2008).

But the epiphany of the face—if we are to do justice to the singularity of a given face—does not adhere to this kind of standard definitional schema. The ethical force of a face is not something constant and universal, bearing the same interruptive power no matter the context in which it confronts us and no matter who the interrupted subject might be. Faces are *always* encountered in particular mediated contexts and against specific socio-cultural horizons. As such, the force of a face can be more or less constrained or blunted depending on the depth and extent of the schemas through which and against which it manifests itself. In the case of human encounters with roadkill, there is an enormous apparatus—both material and ideological—that operates to dull its force. That many people (but, to be sure, not all people) fail to register any call or affect from the bodies of dead animals on our roadways speaks less to some brute fact about their proto-ethical potential and more to the ideological and institutional constraints that encourage and form us not to attend to such bodies.

Consequently, the question of whether roadkill has a face cannot be answered simply by carrying out a more careful phenomenology (as Levinas would have it[3]); instead, it might be necessary for us first to become different kinds of subjects and to prepare ourselves as much as possible for a genuine encounter with roadkill. Stated otherwise, we might be unable to appreciate the ethical significance of roadkill until we become the kinds of subjects who have formed a desire to attend to it and to have carried out such practices of attention for some time. The point here is assuredly not that roadkill will have whatever meaning or value human beings determine it to have, as if we simply impose meanings on an inert world. Rather, the point is that it might be necessary for us to become different kinds of subjects before we can more fully respond to certain bodies and relations that confront us with their uncanny force—especially when we have been so persistently and thoroughly disciplined to be *in*attentive to relations in certain registers of existence. If, then, we are to answer the question, "What *is* roadkill?," perhaps we will first have to undergo a certain subjective de-constitution and re-constitution. For even if (following Levinasian logic) the Other comes to us by "grace," there is often an *askēsis*—a discipline or practice—one needs to undergo in order

[3]When pressed on the question of whether an animal might have a face, Levinas suggests that a "more specific analysis is needed" to answer the question (Levinas 1988, 172).

better to catch sight of and respond to certain Others. Without this critical and reconstructive formation, forces might be felt and Others might become manifest, but only so faintly as to have little or no transformative effect.

Image

As a means of preparing for such an *askēsis*, let us hold the concept of the face in abeyance and consider whether roadkill might better be conceived in relation to the concept of *image* in Maurice Blanchot's sense of the term, a concept that he develops specifically in view of trying to attend to ontological uncanniness. For Blanchot, images work in two ways. On the one hand, we can use images to "humanize" a fundamentally inhuman reality, to render "the formless nothingness ... pleasing and pure" (Blanchot 1982, 255). We accomplish this aim by clothing the world in concepts and meaningful narratives that individuate and substantialize existence. Rather than confronting the sheer uncanniness of neutral being (the *il y a*, to use Blanchot's terminology), we give it contours and forms that aim to strip it of its fundamental alterity. Blanchot suggests, however, that such anthropomorphizing runs up against strict limits when we (i.e., those of us who presumably live in a society that is "right side up"[4]) confront and are confronted by a corpse. The corpse jams our conceptual-ontological machinery, as it is neither a fully living subject nor inert matter. A corpse exists *in* our meaningful world, yet it does not *belong* to it or function according to that world's rhythms. Thus, a cadaver points toward "the possibility of a world behind the world" (Blanchot 1982, 257). In most instances, a corpse cannot be incorporated smoothly and seamlessly into our average, everyday worlds, and yet it *is there* in some sense, it has some force to interrupt. The corpse, then, presents us—beyond our

[4] It is important to note that Blanchot's reflections on the corpse make sense only within the context of a world that is (to use Levinasian terminology) "right side up"—which is to say, a world that accords human bodies some semblance of meaning and value and in which corpses do not pile up unattended on the side of the road. As should be clear, there are situations in which such worlds break down and in which human corpses do pile up, unattended and unmourned. How might consideration of the fate of human bodies in such broken worlds help to illuminate the status of roadkill in worlds that we might otherwise consider to be right side up? This is a question to which I return in the following section in examining Calvin Warren's work.

will—with another sense of an image, an image of a world behind our average, everyday worlds: an image of *the outside*.

If the perspective offered by Blanchot on the human corpse is extended to roadkill, it could be suggested that the general avoidance of serious consideration of this phenomenon indicates a response that involves but goes beyond death denial (a theme to which I will return at more length in Chap. 3 in relation to roadkill art). It is not just that one averts one's gaze and attention from roadkill in order to avoid thinking about mortality; rather, roadkill makes us turn away because it confronts us with the uncanny nature of a mode of existence that is difficult to align with our meaningful projects. The phenomenon of roadkill gestures, on this interpretation, toward *useless* existence, toward the stubborn nature of an "outside" that refuses to go along with the smooth flow of our pragmatic projects. Here, it is sheer existence—"existence without objects" (Bruns 1997, 56), to borrow Gerald Bruns' apt phrase—that haunts us and undercuts our meaningful tasks and goals. From this perspective, roadkill would serve both as evidence that deindividuated existence persists behind our meaningful worlds and as a reminder that we are, at bottom, indistinct from the neutral existence we seek to render into distinct and discrete segments.

Yet, for all of Blanchot's insight here, it does not suffice to reduce roadkill to an uncanny human corpse in a society that is "right side up." For Blanchot's human corpse once inhabited shared meaningful human worlds; it once participated in projects with and alongside other human beings. What makes the human corpse uncanny, according to Blanchot, is that it has left behind such meaningful worlds even as it stubbornly persists in another existential register on their edges. With roadkill, the situation is usually somewhat different. Unless we pause and attend to roadkill with sustained consideration, we will likely not be struck by the fact that the animals littering and haunting our roadsides *once belonged to their own worlds*. Even if one believes (following a certain misreading of Heidegger, and against all evidence to the contrary) that animals are "poor in world," they are far from world-less. The rabbit, the crow, the hummingbird, the raccoon, the possum, the deer—before being struck and killed by vehicles—belonged to rich networks of lives and relations about which most of us are only barely cognizant. To consider the fate of roadkill is thus not simply to encounter an objectless existence that surges up from behind our meaningful worlds, but it is also to catch a glimpse of the innumerable life-worlds that exist within, alongside, and beyond our own (Smith 2009).

Roadkill also differs from the Blanchotian human corpse in that the human corpse is eventually "dealt with" in a society that is right side up and assigned its proper resting place. In this process, the corpse is extracted—for a time, at least—from the flows of life and death and assigned a place—to some extent, at least—within our meaningful worlds. With a "properly" attended corpse, we are spared the difficulty of pondering its ongoing decay or its relation to other beings with whom it might come into contact. With roadkill, by contrast, its relation to the processes of decay and reintegration into trophic chains are often clearly in view.[5] We see a deer carcass slowly being picked over by scavengers; we catch glimpses of a possum day after day that is slowly decomposing in the same spot in the grass on the side of the road; we run over the same flattened squirrel in the road for weeks on end, as it is hollowed out by insects and becomes indiscernible from the pavement underneath. Although the established order tends to direct our attention away from ecological and trophic relations and seeks temporarily to remove (at least certain) human bodies from these cycles of exchange, roadkill is an ongoing reminder that bodies—both animal and human—are not simply part of the *il y a* but are also meat and sustenance for other beings. To attend to roadkill is to be reminded of this basic (and shared) structure of earthly existence.

Terror

Thus far I have suggested that many commonly employed ancient and contemporary theoretical frameworks fail to do justice to the singular and uncanny ontological status of roadkill. But these inadequacies have functioned positively to indicate that not only are alternative frameworks necessary for coming to grips with roadkill, but that there is a corresponding need to undergo a certain transformation in life and thought in order to take up such a project. It would seem, then, that if we wish to know something of what roadkill is, we will have to develop a passion and disposition for attending to the entire arc of life-death in which roadkill participates and to the orders of significance to which it does and does not belong.

To proceed further along this path, perhaps we should take up a perspective in which society is *not* "right side up"—at least for certain

[5] Roadkill removal services are not uncommon in larger cities, but their primary charge is to ensure the smooth and safe flow of automobile traffic—and not to ensure a fitting or respectful burial for dead animals.

inhabitants of the social order—and develop an ontology starting from this point of departure. Indeed, if the concept of anthropocentrism elaborated in the Introduction has hit its target, we should expect there to be a large number of beings—both human and more-than-human—who bear a tenuous relationship to the established socio-ontological order of the human. In his remarkable work, *Ontological Terror: Blackness, Nihilism, and Emancipation*, Calvin Warren adopts something like this starting point in order to explore the status of blacks in an anti-black world.[6] I want to examine Warren's approach at some length here, as I believe it is essential not only for understanding the ontology of blackness but also for gaining insight into the ontological status of other beings, such as road-killed animals, who reside on the underside of the anthropocentric social order.[7] Warren's argument begins from the provocative premise that, if we are to grasp the ontological standing of blacks within the established order, we must do away not only with the notion that things are right side up in the dominant culture for blacks but also with the well-meaning belief that this social order can be reformed such that things will eventually become right side up. In other words, Warren maintains that, as both a practical and an ontological matter, blacks must learn resolutely to refuse the central illusion of the humanist social order, namely, that blacks can ever *be* fully and properly human or that they "can ground existence in the same being of the human" (Warren 2018, 6).

The source of Warren's pessimism about the promise of humanization for blacks stems from his conviction that blacks are introduced into the dominant ontological order not as human *persons* but (in Heideggerian terms) as "available equipment in human form" (6). Given that blacks are framed and positioned essentially as equipment—or what Warren also refers to in more explicitly economic terms as "articles of merchandise" (78, 88)—by the dominant social order, any attempt to transform the social standing of blacks within the terms and conditions of that same order is fated to fail. Thus, Warren suggests that standard appeals to the biological resemblance and continuity among all members of *Homo*

[6] Warren typically refers to "blacks" in the lower case; I follow his practice here when referring to his work.

[7] I should underscore here at the outset of my discussion that marking this adjacency of blacks and animals (which Warren himself also underscores) is by no means intended to denote a simple identity of these groups, either in terms of ontological status or ethico-political standing. The connections here are considerably more fraught and complex, as I hope the following analysis demonstrates.

sapiens will do little to ensure the ontological humanity of blacks. For even if such standing is recognized in certain legal contexts, the possibility of occupying "the human" as an ontological subject position is something that is structurally barred for blacks insofar as the established ontological order of being and meaning is founded on their being equipment and merchandise. This is why Warren insists that the human is "an ontological relation and not a mere legal designation" (88) and that changes at the level of law do not necessarily entail corresponding ontological transformations.

If, then, according to Warren blacks cannot be fitted into the existing socio-ontological order as full and genuine human beings, what *is* their ontological status? Or, as Warren phrases the question: "How do you represent that which lacks a place?" (145). In trying to account for the non-standing of beings who reside outside the socio-ontological order of the human, we encounter the possibility of glimpsing the full depths of the issue that has been haunting us in this chapter. After all, how does one think about the existence of beings who fail to align with the dominant conditions of intelligibility and mattering, especially when "thinking" is often reduced to and equated with extant social and ontological categories? For Warren, the only way to address and push through this critical limit is to acknowledge that this *atopos* zone, this site without a site, is—at least in relation to the dominant social order—a zone of absolute *terror*. That which has been assigned to this place without a place—whether in the form of equipment, or merchandise, or in some other sub- or de-humanized mode—exists without the security of a grounding in the dominant anthropocentric order of meaning. For Warren, whether blacks acknowledge this fact or disavow it, the difficult truth of the ontology of black existence is that it is saturated in terror: "the terror that ontological security is gone, the terror that ethical claims no longer have an anchor, and the terror of inhabiting existence outside the precincts of humanity and its humanism" (4).

Terror must be understood, then, as a byproduct of the established social order—that is to say, it is the status quo itself that renders black existence terror-filled and marks such non-sites beyond its purview in this manner, and it is entirely understandable to desire to exit this terrible non-site and to join the safety and guarantees of the identifiable and secure *topos* of the human and the anthropocentric order of meaning and value. That desire, of course, is the chief affect that animates what Warren refers to as black "humanist" politics. Warren would have blacks ask, though:

What if that humanist desire and its attendant politics is a fool's errand—an ideal that is conceptually, structurally, and ontologically barred from becoming a reality for blacks? Moreover, given Warren's analysis, it is not at all clear that even if gaining such access to the anthropocentric ontological order were a genuine possibility for blacks, it would be desirable to pursue this path. For the "world" that produces such sites of terror is not likely to abandon its logic of splitting the whole of beings into those who matter (the human) and those that do not (blacks and all of the human's Others), inasmuch as the rendering of such a split is the decisive onto-political gesture and structural logic of that world (Agamben 2004).

Rather than endlessly trying to expand, refine, and reform the established order in the direction of humanist justice, Warren suggests that an ontological revolution for blacks is necessary, a revolution that calls for nothing less than the end of "the world"—which is to say, the end of the anti-black, anthropocentric established order that renders blacks as available equipment. But the institution of the end of *this* world does not entail, on Warren's account, an inevitable, easy, and smooth transition to another, better world within the wineskins of the old (as progressives and pragmatists might have it). Rather, to revolt and militate for the end of the world entails, first and foremost, the affirmation of nothingness—of the dread and the terror of the outside of the dominant ontological order—as well as the lucid acknowledgment that any form of life and death worth having must acknowledge this abyss as its (non-)ground. Thus, militating for the end of the world for Warren necessitates a radical exit from the dominant world of meaning and bidding a definitive "adieu to the human" (169) and its onto-political frame, with no nostalgia and no desire for a possible return. Here, in this *other* non-site, where the human is no longer the operative source of order and meaning, human and animal existence are rendered indistinct—and not, Warren suggests, as a triumphant "dethroning of anthropocentrism" (156), but as a clear-sighted recognition that the terror of residing in a liminal relation to the dominant order is shared and distributed among a variety of beings.

If Warren's analysis underscores the terror of blackness in its relation to animality, this conjunction should not be taken to suggest that blacks and animals stand in a simple, isomorphic relation to the human, either normatively or ontologically. Rather, what is indicated by the indistinction of blackness and animality here is the shared non-locus of those who have been consigned to the constitutive outside of the human. Thus, we should not be overly hasty in aligning or reductively identifying the situation of animals with that of blacks—and not just because such alignments are sometimes considered inherently denigratory to blacks (Ko 2019). What is at stake here is trying to catch sight of the ways in which anthropocentrism produces a space both internal and external to "the human" itself, a non-space in which both animality and blackness are situated in related but differing ways. Blackness and animality are thus not to be understood simply as functionally identical instances of the "Other" of the human but are instead singular and always morphing forms of existence that are in the position of being potentially subject to the terror and dread of the anthropocentric ontological order. Stated otherwise, blacks and animals are rendered indistinct within the purview of Warren's framework, not because they are essentially "identical" or "the same," but because they occupy a terror-filled, fraught, and differentially experienced relation to the anthropocentric order of meaning and its concomitant way of life.

We are now better positioned to understand one of the key reasons why the ontologies and frameworks touched on in the first half of the chapter fail to do justice to the ontology of roadkill: namely, they fail to account for ontological terror. They tend to approach existence and its basic structures starting from categories and concepts that emerge from worlds that are salient to all-too-human perspectives and concerns, whether in the form of substances, objects, relations, or subjects. What these frameworks generally fail to do is generate ontological frames from a position that has been structurally barred from the established anthropocentric order of salience. Broaching this limit is the central achievement of Warren's analysis, and the ontological opening he provides for understanding the situation of blacks is equally important for coming to grips with the ontological status of roadkill and other sacrificeable lives. For the animals who become roadkill (along with many other animals and beings of varied sorts) also reside (albeit, differentially) in a site of terror, exposed to the terrible machinations of an anthropocentric form of life become global, outside the guardrails of assurance and registers of salience that structure anthropocentric ontologies.

This non-site of terror analyzed by Warren should be understood, then, as gesturing toward an outside more radical and more expansive than Blanchot's notion of the outside, for Blanchot's outside is framed in relation to a persisting social order that serves as the horizon and frame for grasping the outside of that order. By contrast, the outside at issue in relation to the ontological perspective sketched by Warren is one that is, paradoxically, both generated by the anthropocentric social order while ultimately exceeding and subsuming that same order. To take such an outside as one's point of departure is, thus, to understand the anthropocentric ontological order as merely a partial mapping of the real and as subject to being continually undone by the indistinctions produced by a real that both generates and outstrips its orbit.

Ecofeminist philosopher Val Plumwood explicates this more radical and paradoxical outside from another vantage point of terror in the context of an attack she suffered from a crocodile while out canoeing.[8] In her powerful narrative about the attack, Plumwood explains that while in the jaws of the crocodile and being subjected to repeated underwater death rolls, her usual, all-too-human ontological frame is entirely shattered. Accustomed to seeing the world from the inside (i.e., from the perspective of an individual human being with subjective interiority and immersed largely in a world of shared interhuman sociality), the prospect of being eaten by another animal forcibly tears her from that perspective and opens her onto an outside that is radically foreign and indifferent:

> In that flash, when my consciousness had to know the bitter certainty of its end, I glimpsed the world for the first time "from the outside," as a world no longer my own, an unrecognizable bleak landscape composed of raw necessity, that would go on without me, indifferent to my will and struggle, to my life or death. (Plumwood 2000, 58)

The terror that Plumwood feels when the assurances of anthropocentric security slip introduce her to a zone that Warren suggests blacks inhabit—again, whether they acknowledge or disavow it—on an ongoing basis with regard to the dominant order. Here, in the outside, there is existence (living and dying) without existence (socially intelligible lives and deaths).

Event

Yet, it is equally important to emphasize that this outside is not *exclusively* a zone of terror or lacking in meaning; to assume as much would be to grant a totalizing power to the established anthropocentric order. Before

[8] Plumwood discusses this attack in several publications; I draw here primarily from Plumwood (2000). I have examined this piece in more detail in Calarco (2022).

and beyond the ontological terror associated with the human, there lies a world—nay, many worlds—that are more than, richer than the anti-black, anti-animal, all-too-human world. Toward the end of his work, Warren refers to these other worlds found among and on the obverse of terror as zones of "majesty" (171) that exceed and precede dominant coordinates. We might say, then, that beyond the interests and purview of the human and circulating through and outside the limits of its reign of terror, there are innumerable worlds and zones of more-than-human and inhuman existence laden with counter-possibilities and potentialities. In a series of reflections on blackness and animality that complement Warren's analysis in fecund ways, Joshua Bennett describes such potentialities and worlds as a "gift of blackness," where blackness is understood as the ambiguous and paradoxical ontological situation within which black lives and deaths circulate. For Bennett, these other worlds and ontological zones represent "the great chain of being come undone, life itself unfettered and moving in all directions" (Bennett 2020, 4). Here is opened a "window into the worlds that thrive at the underside of modernity," worlds that are "nothing short of another cosmos" (Bennett 2020, 5). In this sense, the ontological revolution needed to displace the dominant order of the human does not function according to a religious logic of the *saeculum* or the *eschaton*. We need not await it, nor need we create the conditions for its advent. These other worlds, notes Bennett, are "already here, already in the works, already waiting for us in the wild" (Bennett 2020, 5). The critical, post-anthropocentric task is thus to learn critically to occupy this uncanny site and strive to be done with, truly bid adieu to, the human.

Similarly, Plumwood's experience of a radical, non-anthropocentric outside cannot be reduced exclusively to an experience of terror and dread. The more-than-human, inhuman, and radically indifferent outside she catches sight of during the attack can also be approached—even affirmed—from other, non-anthropocentric angles. In this vein, where Warren writes of approaching existence outside the human in terms of its "majesty" and Bennett refers to this non-site as "another cosmos," Plumwood refers to the outside as a gift-giving source of profound wonder. She notes that, after her attack, the wonder of seeing the world from the outside

> has never entirely left me. For the first year, the experience of existence as an unexpected blessing cast a golden glow over my life, despite the injuries and

> the pain. The glow has slowly faded, but some of the gratitude for life it left will always be there, even if I remain unsure whom I should thank. The gift of gratitude came from that searing flash of near-death knowledge, a glimpse "from the outside" of the alien, incomprehensible world in which the narrative of self has ended. (Plumwood 2000, 59)

I would suggest that what is at stake with this rhetoric of the radical outside, majesty, another cosmos, wonder, and so on, is the simple recognition of the fact that world *is given*, that it is there before us—all of us, both those who accede to the subject position of the human and those who are structurally barred from that order. The radical outside is not waiting for us to grant it presence or to bring it to light but is unfolding with and without us in that non-locus of the wondrous and monstrous majesty of existence. To catch sight of the outside is to see and feel—with brute existential force—*that* there is something rather than nothing (in the sense that Martin Heidegger (2014) grants to this phrase). This is the shock that forces us to think, that insists we respond, and that calls us to learn to live differently. The dominant anthropocentric order is founded on a refusal of this givenness, a refusal to think the outside and a sustained effort to reduce the world to its meaning and its terror. The task of countering anthropocentrism begins, then, from an acknowledgment of the ontological terror associated with the human as well as an affirmation of the givenness of a world that exceeds its ontological orbit and its concomitant social order.

If we are to undertake something like an ontology of and reflective engagement with roadkill, I would suggest that it be approached from this double angle, namely, that of terror and its majestic, wondrous, and ambiguous obverse. I seek to enact something like this approach in the remaining chapters of this book. The crucial, supporting point I seek to make in these pages is that grasping something of the significance of roadkill requires the development of a counter-disposition and a taste for attending to this outside and these other worlds, a decelerative posture (Vannini 2014), that avoids being overly hasty in assigning a meaning to roadkill and instead remains patiently with it and allows it to work us over. So rather than assigning a familiar ontological category to roadkill, we might consider its uncanniness as signaling its status as an *event*—that is, as something for which we lack adequate categories and for which concepts and responses must be created anew.[9] To approach roadkill in this manner would involve not simply assigning it to one or more extant ontological categories or frameworks but to recognize its givenness, and in so doing, allowing it to affect our thought and our practice and then seeking to respond to it.

I will have more to say about this notion of roadkill as an event in the chapters that follow. For now, though, I should note that the most basic valence attached to events—and one that will be of central importance to my analysis—is that they mark ruptures in the established order of meaning. In the introduction, I postulated that anthropocentrism and hyperautomobility are among the most important systems structuring the established order in the contemporary age. To suggest that roadkill is an event for the present is thus to understand it as an event in relation to that social order of meaning. In underscoring the relativity of this rupture, though, I am not arguing that the significance of roadkill is *entirely* relative to this established order. To the contrary, and following the line of argument developed in this chapter, to encounter roadkill as an event implies that it cannot be made full sense of using categories, concepts, or institutional arrangements from or related to the current anthropo-hyper-auto-mobile order. In approaching roadkill as an event, we are thus challenged to see it as something that exceeds our purview and that solicits our standard ontological frameworks.[10] To be sure, our concepts and categories help partially to illuminate roadkill: we see this deer struck by a semi-truck, or that badger hit by a car, or these crows scavenging a roadkilled mouse lying on the pavement. All too often, though, our standard frames and categories allow us to pass by and over the brute fact—the givenness—of *this* particular roadkilled animal in its uncanny ontological form, the world that was shattered with the end of its life, and the new worlds that are forming with and through its death. Encountered as an event, roadkill forces us to *respond*—that is, to create concepts and practices, to

[9] The concept of event has a wide circulation in contemporary theory and philosophy. It plays a significant role in the philosophy of Martin Heidegger, Jean-François Lyotard, Jacques Derrida, Emmanuel Levinas, Gilles Deleuze, and Alain Badiou, to name a few of the more prominent figures. In proposing that roadkill be conceived of as an event, this body of work will certainly be relevant to my concerns, but I depart from the standard uses of this term in ways that will be apparent in the pages that follow.

[10] I use the term solicit here in the etymological sense emphasized in Derrida's work, in which to solicit means to make something "shake as a whole, to make [it] tremble in [its] entirety" (Derrida 1982, 21).

think, be, and move otherwise—in order to do justice to these more complex realities that are occluded by the dominant linguistic and socio-ontological order. Developing the chief elements of such a response is the central task of the chapters that follow.

References

Agamben, Giorgio. 2004. *The Open: Man and Animal.* Translated by Kevin Attell. Stanford, CA: Stanford University Press.

Aristotle. 1987. Categories. In *A New Aristotle Reader*, ed. J.L. Ackrill. Princeton, NJ: Princeton University Press.

Bennett, Joshua. 2020. *Being Property Once Myself: Blackness and the End of Man.* Cambridge, MA: Harvard University Press.

Blanchot, Maurice. 1982. *The Space of Literature.* Translated by Ann Smock. Lincoln, NE: University of Nebraska Press.

Bruns, Gerald L. 1997. *Maurice Blanchot: The Refusal of Philosophy.* Baltimore, MD: Johns Hopkins University Press.

Calarco, Matthew. 2022. *The Boundaries of Human Nature: The Philosophical Animal from Plato to Haraway.* New York: Columbia University Press.

Derrida, Jacques. 1982. *Margins of Philosophy.* Translated by Alan Bass. Chicago: University of Chicago Press.

Desmond, Jane. 2016. *Displaying Death and Animating Life: Human-Animal Relations in Art, Science, and Everyday Life.* Chicago: University of Chicago Press.

Heidegger, Martin. 2014. *Introduction to Metaphysics.* Translated by Gregory Fried and Richard Polt. New Haven: Yale University Press.

Ko, Aph. 2019. *Racism as Zoological Witchcraft: A Guide for Getting Out.* Brooklyn, NY: Lantern Books.

Levinas, Emmanuel. 1969. *Totality and Infinity: An Essay on Exteriority.* Translated by Alphonso Lingis. Pittsburgh, PA: Duquesne University Press.

———. 1974. Philosophy and the Idea of Infinity. In *Collected Philosophical Papers*, 47–60. The Hague: Martinus Nijhoff.

———. 1988. The Paradox of Morality: An Interview with Emmanuel Levinas. In *The Provocation of Levinas: Rethinking the Other*, ed. Robert Bernasconi and David Wood, 168–180. London: Routledge.

Lulka, David. 2008. The Intimate Hybridity of Roadkill: A Beckettian View of Dismay and Persistence. *Emotion, Space and Society* 1 (1): 38–47.

Plato. 1997. Sophist. In *Plato: Complete Works*, ed. John M. Cooper. Indianapolis, IN: Hackett.

Plumwood, Val. 2000. Being Prey. *Utne Reader* 100: 56–61.

Smith, Mick. 2009. Road Kill: Remembering What Is Left in Our Encounters with Animals. In *Emotion, Place and Culture*, ed. Mick Smith, Joyce Davidson, Laura Cameron, and Liz Bondi, 21–34. Burlington, VT: Ashgate.
Vannini, Phillip. 2014. Slowness and Deceleration. In *The Routledge Handbook of Mobilities*, ed. Peter Adey, Mimi Sheller, David Bissell, Kevin Hannam, and Peter Merriman, 116–124. New York: Routledge.
Warren, Calvin L. 2018. *Ontological Terror: Blackness, Nihilism, and Emancipation*. Durham, NC: Duke University Press.

CHAPTER 3

Roadkill and Other Sacrificeable Lives

Abstract This chapter scrutinizes the "pragmatic attitude" that treats animals as mere obstacles to the smooth flow of circuits of mobility. At the heart of this attitude, it is argued, is a logic of sacrifice that structurally excludes animals and others deemed non- and sub-human from the circle of social and normative consideration. Critical responses and alternatives to this logic are canvassed under three different rubrics: liberal humanism, zoo-pessimism, and inhumanism. The chapter concludes with a consideration of how contemporary roadkill artists endeavor to challenge the logic of sacrifice and how their work aligns with key elements of the inhumanist perspective.

Keywords Grievable • Killable • Sacrificeable • Inhumanism • Roadkill art

Obstructions

In *Eclipse of Reason*, Max Horkheimer offers an anecdote to illustrate what he labels "modern insensitivity to nature" (Horkheimer 2004, 71). In his discussion of the logic of domination as applied to the natural world, Horkheimer mentions how the landing of planes in Africa at the time (which presumably occurred in a region that had not had previous

M. Calarco, *Reflections on Roadkill between Mobility Studies and Animal Studies*, https://doi.org/10.1007/978-3-031-30578-8_3

substantial air travel infrastructure in place) was seen as often being "hampered by herds of elephants and other beasts" (Horkheimer 2004, 71). That the air travel infrastructure and landings were at odds with the welfare of elephants and other animals did not, Horkheimer notes, lead to a rethinking or curbing of such practices; instead, the animals themselves were viewed as constituting the problem. In this situation animals were, in Horkheimer's words, "considered simply as obstructers of traffic" (Horkheimer 2004, 71) and their lives and deaths were understood to have no significance or importance in their own right.

Horkheimer links this insensitivity to the lives and deaths of animals to a "pragmatic attitude" (Horkheimer 2004, 71) that he argues is operative throughout the history of Western thought and culture up to the present. My aim in this chapter is to probe further into this "pragmatic attitude," specifically in relation to the problem of roadkill. I do so through an examination of what I call a *logic of sacrifice* that structures and permeates the pragmatics of mobility. I explicate this logic by way of an engagement with the writings of Judith Butler on grievability, Donna Haraway on killability, and Jacques Derrida on sacrificeability; these analyses are each in their own way concerned to demonstrate that the dominant social order is predicated on the structural sacrifice and exclusion of certain beings from the circle of social and normative consideration. Extending these analyses, I seek to explore how the logic of sacrifice functions in the context of modern systems of mobility to frame animals as beings who can be routinely sacrificed in the service of the established social order of mobility. I then consider possible critical responses and alternatives to this logic under three different rubrics: liberal humanism, zoo-pessimism, and inhumanism. I close the chapter by examining how certain contemporary roadkill artists endeavor to challenge the logic of sacrifice and how their work aligns with key elements of the inhumanist perspective.

Logics of Sacrifice

Judith Butler's *Frames of War* (Butler 2009) is the *locus classicus* for the contemporary dialogue that has grown up around the concept of *grievability*. In this work, Butler develops a richly layered notion of grievability in view of discerning how some beings are framed as mattering while others are made not to matter—which is to say, how some beings bear social significance while others lack it. She suggests that one way to understand a particular being's social significance and standing is to consider whether it is "grievable" by the dominant standards of a given social order. To be grievable on this account is to have a death that is considered worthy of mourning, a death that has some significance and that is marked as mattering in various ways by the moral community. So, in response to the death of human beings who are seen as grievable, a community might participate in certain funeral rites, orations, days of remembrance, and other such ceremonies that mark that person as being significant. Butler proposes that grievability of this sort is "a presupposition for [a] *life* that matters" and "a condition of a life's emergence and sustenance" (Butler 2009, 14–15). Thus, for a life to be livable, the death of that life must at the same time be framed as grievable. Butler writes:

> Without grievability, there is no life, or, rather, there is something living that is other than life. Instead, "there is a life that will never have been lived," sustained by no regard, no testimony, and ungrieved when lost. Grievability precedes and makes possible the apprehension of the living being as living. (Butler 2009, 15)

That there are lives which amount to something less than and "other than life" speaks to the fact that grievability is differentially allocated by the established order (Butler 2009, 182). Among those who are included in the moral community, grievability is dispensed in varying degrees, and for those who are excluded from consideration, grievability is often entirely denied. In line with the analysis of anthropocentrism in Chap. 1, Butler clearly recognizes that the complex borderlines of grievability run in concert with the contours of the subject position of the human (Butler 2009, 76). Those who are genuinely grievable, she suggests, are those deemed to be fully and properly human, and the degree to which a given being is understood to participate in and depart from that subject position functions to determine the relative worth of its life and death.

So, how to name and think more carefully about those beings who are *denied* grievability? Butler uses the term "lose-able" for this task, a concept that speaks to the logic of sacrifice I will be elaborating in what follows. In referring to beings and groups who have been exiled from the circle of social consideration, Butler notes:

> Such populations are "lose-able," or can be forfeited, precisely because they are framed as being already lost or forfeited; they are cast as threats to human life as we know it rather than as living populations in need of protection

> from illegitimate state violence, famine, or pandemics. Consequently, when such lives are lost they are not grievable, since, in the twisted logic that rationalizes their death, the loss of such populations is deemed necessary to protect the lives of "the living." (Butler 2009, 31)

Sociologist David Redmalm offers a helpful rubric for understanding how beings become lose-able in the sense articulated by Butler. On Redmalm's account, the individual deemed not worthy of grieving is de-singularized and "framed as replaceable" (Redmalm 2015, 23). In addition, the death of a lose-able individual is understood to have no transformative force (to recall the discussion of substances ontologies in Chap. 2) and to belong to a predictable and insignificant course of events. Finally, the lose-able individual's death is characterized as one that is not shared by those who matter. Their mortality and finitude do not denote "a common state of precariousness" (Redmalm 2015, 23) but belong to a different order of death from which members of the dominant social order are believed to be exempt.

In a gesture that resonates with my central concern in this chapter, Redmalm goes on to take up the question of how individuals deal with the death of animal pets. As he suggests, in the ambiguous and fraught process of grieving for pets, the grieving person is often times trying in some way to shift the status of that individual animal from being lose-able to being grievable, against the backdrop of social conditions where grievability is largely coterminous with the human (Redmalm 2015, 23). Although Butler herself does not focus in detail on the grievability and lose-ability of pets or animals more generally, she is nevertheless cognizant that her discourse on grief has the potential to contest the anthropocentrism of dominant attitudes toward animals and other beings who circulate on and outside the margins of the human. In this vein, Butler does not rest content with the facile assumption that the differential allocation of grievability can be solved by an appeal to straightforwardly humanist premises or to a common human nature, for such gestures simply reinforce anthropocentric dogmas. Thus, in reflecting on the limits of a universalist humanism in thinking about grievability, Butler asks:

> How does one object to human suffering without perpetuating a form of anthropocentrism that has so readily been used for destructive purposes? Do I need to make plain in what I consider the human to consist? I propose that we consider the way "the human" works as a differential norm: Let us think

> of the human as a value and a morphology that may be allocated and retracted, aggrandized, personified, degraded and disavowed, elevated and affirmed. The norm continues to produce the nearly impossible paradox of a human who is no human, or of the human who effaces the human as it is otherwise known. Wherever there is the human, there is the inhuman. (Butler 2009, 76)

I will return to the concept of the inhuman later in this chapter. For the moment, though, I want to suggest that, despite Butler's awareness of the problems associated with naïve anthropocentrism, there are certain limits to her approach when it is applied directly to roadkilled animals. For there is a significant difference between (1) attending to the general ways in which the concept of the human produces a differentially populated constitutive outside of de-humanized and sub-humanized others (which is characteristic of Butler's approach) and (2) understanding the specific ways in which discourses and practices related to the grievability and lose-ability of life and death produce and normalize the routine sacrifice of roadkill (which is part of my task in this chapter). Although these two projects are logically related, there is precious little direct attention to animals themselves (whether living or dead) in Butler's work or to the sorts of questions posed by Redmalm and other animal studies scholars about the relative lose-ability and grievability of animals (Stanescu 2012). In order to address this limitation, it will be helpful to examine how Donna Haraway and Jacques Derrida—two authors who have written at length on animals themselves—frame the question of animal life and death in relation to the established anthropocentric order.

For Haraway, the question concerning animal mattering is one of whether animals are made *killable*, which is to say, whether they can be killed with moral impunity and whether they are understood ontologically as being no more than entities to be used and discarded (Haraway 2008). The notion of killability is essential for understanding Haraway's approach, for she is keen to distance herself from certain animal rights and animal liberation frameworks that (on her account) try to remove animals *tout court* from the realms of death, predation, and other modes of violence. Haraway's thought presumes, by contrast, that life is inherently violent and involves ubiquitous killing and consumption. There is, she argues, no way to avoid these interlaced cycles of life and death, of giving birth and bringing death. So, rather than seeking to establish bulwarks that protect animals from any and all forms of violence and death, Haraway argues instead that we need to guard against "exterminism" and reducing animals and others to the realm of beings that are "merely" killable (Haraway 2008, 78–80). In Haraway's words, "it is not killing that gets us into exterminism, but making beings killable" (Haraway 2008, 80). If, then, we are not entirely able to avoid killing, the problem becomes one of learning to

> live responsibly within the multiplicitous necessity and labor of killing, so as to be in the open, in quest of the capacity to respond in relentless historical, nonteleological, multispecies contingency. Perhaps the commandment ["Thou shalt not kill"] should read, "Thou shalt not make killable." (Haraway 2008, 80)

Haraway's point about the "necessity and labor of killing" is apt, and one that is all too easy to disavow when considering roadkill in particular, where the temptation might be to try effectively to eliminate roadkill entirely through certain activist and policy measures. Animals who are not killed by cars will, however, be killed or die in some other way—and in a way that might be worse than death by car, plane, or train. So, our analysis should bear in mind Haraway's point about the necessity of killing and be on alert for the desire to establish a good conscience and clean hands with respect to animal death.

I would suggest, though, that Haraway's approach to killability—for all of its evident insight—is nevertheless problematic inasmuch as it tends to allow the irreducibility of violence and the necessity of killing to occlude the possibility of making radical changes in certain behaviors and institutions that can partially ameliorate useless suffering and premature death. For example, Haraway will consider the possibility of eating meat that has been raised under better welfare conditions or improving the conditions of experimental animals, but she tends to dismiss veganism or the abolition of invasive animal experimentation as naïve ideals that turn farmed and experimental animals into living museum pieces (Williams 2010; Franklin 2017). Her stance seems to presume that if these animals are not farmed for food or used for the advancement of science, they will not have a life worth living. It is better, then, she argues to maintain these historical relations that involve killing but that also allow for the ongoing possibility of living. We could insist in response that to struggle to free animals from such forms of confinement and commodification and toward other potentialities is not so much a matter of denying their finitude or mortality or keeping them protected from violence and on display in a museum; rather, such efforts are an attempt to free animals for a more worthwhile life—and death. The same should be true of the struggle to reduce the number of roadkilled animals. The point of such work ought not be somehow to remove animals from the condition of being embodied and mortal but should instead seek to challenge the multiple ways in which modern systems of mobility cause useless suffering and deaths, actions that are themselves ethically problematic and that also create tears in the fabric of life that tend to rip and contribute to other unnecessary cuts and crises.

In brief, then, what goes missing from Haraway's analysis of many (but certainly not all) human-animal relations is consideration of the possibility that the established order is susceptible in many instances to being fundamentally contested and radically changed in the direction of maximal respect for animals. Moreover, it should be noted that the making killable of animal life that Haraway seeks to contest is not an unfortunate side-effect of many human-animal relations but is functionally and structurally integral to many of those relations and their institutional settings—which means that contesting the killability of animals often means working radically to curtail and even abolish certain practices. In institutions like factory farming and invasive experimentation, animals are ontologically framed as equipment and commodities (to recall Calvin Warren's analysis of the ontology of black existence from Chap. 2) to be reared and killed for human use, and this frame remains in place even in "free-range" and "green" alternatives to factory farming and in the supposedly "less cruel" alternatives to invasive experimentation (which Haraway tends to favor). Contemporary systems of mobility are no different. The ongoing killing of animals, the fragmentation of animal habitat, the degradation of ecosystems, and myriad other problems that attend the institution and maintenance of circuits of mobility are not accidents of the system but are fundamental to its design, and many of the "solutions" that seek to address roadkill and other harms to the more-than-human world do nothing to challenge the fundamental assumptions that undergird that system.

We might say, then, that the lose-ability and killability of animal life characteristic of many contemporary modes of human-animal interactions are ultimately predicated on the uncritical *sacrifice* of animal life. Phrased otherwise, animals and more-than-human others of many kinds are routinely harmed and killed in order to maintain the dominant way of life. What is more, this way of life—animated by its anthropocentric, hyperautomobile soul—takes itself to be non-negotiable. It allows for minor reforms at the structural level and sometimes seeks to cloak the cruelty of its veneer to some degree, but as a matter of principle it structurally leaves open a space for what Jacques Derrida refers to as sacrifice in the form of the "noncriminal putting to death" of animal life (Derrida and Nancy 1991, 112). This is why Derrida can write, in a way that complements Haraway's insights but simultaneously departs from them, that the commandment "Thou shalt not kill" is never extended by the dominant culture to the "living in general" (Derrida and Nancy 1991, 112).

In this way, Derrida—much like Haraway—underscores the fact that animals are rendered killable by the established order. Derrida emphasizes, however, the inverse point as well—namely, that such killability presupposes a group of beings (namely, those individuals who are taken to be paradigm instances of the human) who render others killable but who exempt themselves from that status. Derrida refers to this gesture of designating others as killable while exempting oneself and one's confederates from this fate as a "sacrificial structure" (Derrida and Nancy 1991, 112). He argues that even the most insightful critics of traditional humanism rarely, if ever, call this sacrificial structure into question—which is to say, these critics of humanism fail to "*sacrifice sacrifice*" and thereby remain dogmatically committed to a certain anthropocentrism (Derrida and Nancy 1991, 112). Following Derrida, we can observe that, despite the proliferation of criticisms of and alternatives to humanism in our own time—from antihumanism to transhumanism to posthumanism—this logic of sacrifice still forms the invisible contours of thought and practice. The standard alternatives to humanism might include a consideration of reforms to this logic, but the logic itself and the broader anthropocentric order to which it belongs lie largely beyond the range of critique.

Liberal Humanism, Zoo-Pessimism, Inhumanism

The discourses I have analyzed thus far in this chapter concerning loseability, killability, and sacrificeability all disclose a structural issue with the dominant social order: namely, that it functions to obscure, minimize, and ultimately justify the sacrifice of countless lives—both human and more-than-human—in the service of constituting and sustaining the dominant way of life. In relation to roadkill in particular, this logic is deployed to justify the ever-increasing presence of automobiles, the continuing expansion of roads and other circuits of mobility, and the minimizing of the violence, harm, and disruption of subjective, social, and ecological relations that circulate through and among animals. Now, even if it is agreed that the field of animal studies is premised on a general rejection of the marginalization and sacrificeability of animal life, it is far less clear what such a rejection involves or what a genuine alternative to this logic might be. Let us, then, sketch a series of possible responses to the logic of sacrificeability from a generally pro-animal perspective and do so specifically in view of the problem of how to address the problem of roadkill. I consider here three different approaches using the rubrics of liberal humanism, zoo-pessimism, and inhumanism.

Consider, first, a *liberal humanist* approach to the problem of sacrificeability. For the liberal humanist, the overarching aim of politics and normative theory is to recast who does and does not count, using the human as the measure. Typically, liberal humanism has limited itself to the project of extending consideration to human beings who have previously been excluded from consideration (people of color, women, people of alternative sexual orientations, and so on), but, as pro-animal advocates have pointed out, there is no reason to restrict such extensionist logic to paradigm members of the biological species of human beings. If the primary criteria for inclusion in the liberal political sphere are subjectivity, consciousness, sentience, autonomy, and other such characteristics, then what it means to be "human" in the political and normative sense is not coterminous with the biological species *Homo sapiens*—for, we find these traits to a greater or lesser degree, pro-animal advocates argue, among animal species of various sorts. Consequently, the reasoning goes, animals displaying these characteristics can be excluded from consideration only by illogical and unjust means. For the pro-animal advocate, strict logic requires that the lives and deaths of animals be seen as having just as much worth and being just as grievable as those of human beings. Human rights, on this account, are thus not exclusively human but belong to a variety of animals as well (Cavalieri 2001). Just as rights are intended to provide (relatively) absolute bulwarks against violations of life, bodily integrity, and autonomy for human beings, giving animals similar rights (it is argued) will prevent them from suffering similar forms of violence.

If we consider roadkill from this perspective, we would assume that critical analysis of systems of mobility would figure centrally in the animal rights discourse and activism that issues from liberal humanist commitments. For, depending on the figures used, roadkill is one of the leading causes of animal injury and death in many advanced industrial societies. By standard estimates, more than one million animals are killed per day in the

United States by cars alone (this figure does not include rail or air or marine traffic; for various estimates, see: Hill et al. 2019; Schwartz et al. 2020; Seiler and Helldin 2006). Depending on the estimate, roadkill numbers approximate and possibly exceed the number of animals killed by such practices as invasive experimentation or legal hunting (ubiquitous topics in the animal rights literature and central issues in animal rights activism). Yet, a relative silence reigns around the issue of roadkill in mainstream forms of pro-animal discourse and activism influenced by liberal humanism. Where we might expect calls for a radical reduction in driving and a complete reconsideration of urban and rural mobility infrastructure, we tend to find among mainstream animal rights and welfare organizations little more than calls for more wildlife crossings, lower speed limits for vehicles, and similar reforms (see, e.g.: People for the Ethical Treatment of Animals n.d.; In Defense of Animals n.d.).[1]

One of the reasons for this relative silence on the issue might be that there is no meaningful consumer-based solution to the problem of roadkill. As just noted, the guiding ideal behind liberal humanist versions of animal politics is that animals should, for the most part, be "lifted up" out of the realm of sacrificeability and be provided the same kinds of strict protections that paradigm instances of the human are understood to bear. The chief strategy for achieving this goal in mainstream animal advocacy circles has generally been consumerist in orientation—which is to say, mainstream animal politics is often premised on having individuals purchase and consume alternative products that avoid using animals in instrumental ways. So, rather than eating animal-based meat, we are encouraged to consume plant-based meat alternatives; rather than purchasing cosmetics tested on animals, we are counseled to buy "cruelty-free" products, and so on. To be sure, there are certain practices from which we are encouraged to abstain *tout court* (e.g., the exploitation of animals in the pet and entertainment industries), but the central message of mainstream animal rights and animal welfare organizations is that no fundamental changes are required in our present (which is to say, consumerist) way of life in order for animals to be granted full ethical and legal standing. It is assumed that we can, for the most part, purchase our way out of injustice.

As Dennis Soron has persuasively argued, however, the kinds of ethico-political intervention required for addressing the problem of roadkill are rather different from these sorts of strategies and tactics (Soron 2008).

[1] These and other mitigation strategies will be discussed at more length in Chap. 5.

While it might be possible to limit and almost eliminate the direct consumption of animal products from one's diet, there is no similar abstention possible in regard to the use of structures of mobility for most individuals, as these circuits and modes of transport permeate every aspect of our lives, in both direct and indirect forms. Nor are there any simple, ready-made consumer substitutes (such as, say, a "cruelty-free" automobile) that avoid the harms done to animals by our hypermobile way of life (although there are certainly *better* modes of mobility possible than the modes currently dominant, a topic I take up in Chap. 5). In short, addressing the issue of roadkill entails a full reconsideration of dominant structures of mobility across a variety of registers, a reconsideration that cannot help but to call fundamentally into question our consumerist and hypermobile form of life. Of course, most mainstream activists, theorists, and organizations do not want to preach this sort of message, for fear of being labeled as fringe or unpragmatic. While such hesitance is understandable, the fact remains that we have collectively built a mobility infrastructure and form of life that can only be made (even partially) just by being (for the most part) dismantled and reconstituted from the ground up.

It might be suggested that another reason the liberal humanist approach runs into an intractable limit with regard to roadkill is because it starts from an overly optimistic assessment of the potential for reform of human-animal relations within the current social order. Rather than assuming that animals can be fitted over time into the dominant normative (ethical, legal, and political) order, it might be more fairly argued (the critic could suggest) that the established order cannot be genuinely reformed with regard to animals. Animals enter this order in various ways: as pets, meat, laborers, clothing, pests, entertainment, resources—in short, as usable, lose-able, killable, and sacrificeable commodities and equipment (to recall Calvin Warren's terms)—but never as beings who exist in and for themselves, as independent of the (lack of) meaning granted to them by the dominant human culture, and almost never as beings with their own significant worlds, relations, lives, and deaths. Our systems of mobility, it could be argued, cannot be reformed in such a way as to recognize the singularity and alterity of animals without failing to exist; rather, the very nature of those systems is founded on the de-singularization of animal life and requires its ongoing sacrifice for continued operation and expansion.

Forthright recognition of this dominant anthropocentric onto-logic concerning animal being might lead us to a different, less optimistic conclusion about addressing the general sacrificeability of animals within the current coordinates of the status quo. Let us refer to an approach based on such recognition as *zoo-pessimism*. For the zoo-pessimist, it is necessary to begin with the admission that animals—contra the liberal humanist approach—are never going to be raised up fully into the sphere of the human, are never going to be granted human rights, and will always be seen as sub-human or less-than-human within the current social order. In the dominant cultural context, animal life is and remains fundamentally lose-able, even if the death of the occasional individual animal (typically, a charismatic animal) might be considered grievable. Animals are and always will be framed, the zoo-pessimist argues, in such a way as to render them killable by human subjects with impunity. Animals are and will continue to be sacrificed by the billions so that our way of life can be maintained. With regard to roadkill in particular, the pessimist would counsel against the naïve belief that mobility infrastructure will ever be fundamentally transformed in a pro-animal direction by the dominant culture. At best, roadkill numbers might be lowered through reducing speed limits and constructing wildlife crossings, but the telos of such policy and infrastructure initiatives is and always will be human wellbeing, whether in view of improving the welfare of human drivers who might be injured in human-animal collisions or in the form of maintaining the enjoyment human beings gain from having ongoing access to wilderness areas richly populated by fauna.

The zoo-pessimist is thus likely to see animal welfare issues in general and the problem of roadkill in particular as intractable problems and as a cause for despair. The minor reforms to mobility infrastructure and the expansion of wildlife crossings and other such strategies celebrated by mainstream pro-animal organizations are bound to appear to the zoo-pessimist as mere band-aids applied to a massive and persistent wound. Moreover, the zoo-pessimist would be disinclined to believe that other, better solutions to the problem of roadkill are possible on the collective level. The social order, it would be argued, is founded on anti-animal, anthropocentric premises that cannot be easily displaced. The best that can be hoped for is that conscientious individuals might adopt an ethic of limiting or eliminating driving and try to remove themselves from the dominant circuits of mobility as much as is reasonably possible. We are caught up in a set of structures, the zoo-pessimist maintains, that has taken on a life of its own, and there is limited agency possible under such conditions, either for human beings or for animals. The zoo-pessimist might also remind us that current forms of hypermobility are only the latest manifestation of an anthropocentrism that has persisted in Western and many other cultures for millennia. Horkheimer's comments on pragmatic attitudes toward animals apply to ages much earlier than our own. Animals have long been viewed as "mere obstructers of traffic" in one way or another. The only genuine hope for a better life for animals, it might be maintained, is the decay and eventual death of large-scale human societies.

If we assess the relative probity of the liberal humanist and zoo-pessimist analyses of the situation concerning the sacrificeability of animals and roadkill, I would suggest that zoo-pessimism is much nearer the mark about the prospects for fundamental change within the coordinates of the established order. The oft-repeated dogma among theorists and activists that we can gradually reform our way to genuine rights for animals within present economic and legal frameworks is clearly no longer tenable, given the meager success this strategy has had over the past several decades. In saying this, I am not suggesting of course that all such efforts at achieving rights or reform are pointless, only that it would be naïve to believe that animals can somehow be fitted over time in a full and genuine way into the liberal humanist social and political order. The zoo-pessimist is right, I think, to be skeptical of this optimistic vision. To extend rights to animals and to fundamentally change the anthropocentrism and hypermobility of the dominant culture cannot be accomplished within the wineskins of the present order; the implementation of such changes would certainly cause those skins to burst—or, to stay with the metaphor, require an entirely different sort of skins.

If zoo-pessimism is helpful for pinpointing the limitations of the optimistic reformism of the liberal humanist approach, it nevertheless bears a serious, critical limitation of its own. In particular, zoo-pessimism shares with liberal humanism a totalizing view of the established anthropocentric order. For, even though it is crucial to mark the ways in which the dominant ontological order recurrently and differentially excludes those beings who are deemed to be less-than-human, it is equally essential to attend to the ways in which alternative registers of meaning, value, and relation circulate through, around, and outside those zones (to reiterate a point we encountered in the work of Calvin Warren, Joshua Bennett, and Val Plumwood in Chap. 2). The established order is but one (admittedly powerful and consequential) order among many. If there is any hope of shrinking the impact of that order (for seeking its utter elimination in the near future would be naïve), such a gesture must ground itself in recognition and affirmation of the rich variety of outsides to that order that are already present.[2]

Let us name an approach grounded on the recognition and affirmation of such outsides *inhumanism*. Beyond liberal humanism (which seeks to raise up animals into the protected realm of the human) and beyond zoo-pessimism (which seeks to demonstrate the dead end of pursuing liberal humanist solutions for animal issues), the inhumanist approach proceeds from the premise that the other worlds, values, relations, and ways of being that constitute genuine alternatives to anthropocentrism are already to some extent present and make their presence known in a variety of ways. The task we face, according to inhumanism, is not so much that of creating such alternatives from scratch; rather, it is one of attending to presences that are not necessarily intelligible in the terms offered by the established order while simultaneously attenuating that order's realm of influence. In other words, the goal envisioned by the inhumanist approach is one of learning to perceive, think, attend to, and value life and death in a fundamentally non-anthropocentric manner.

The most proximate source for this sense of the term inhumanism derives from the poet Robinson Jeffers, who uses it to name the underlying philosophical vision animating his work. Jeffers's mature poetry is aimed throughout at encouraging readers to "turn outward" (Jeffers 1988–2001, 2: 418) and move beyond the human narcissism characteristic of so much of contemporary culture. For Jeffers, this narcissism and anthropocentric myopia regarding the locus of meaning and value constitute the chief sources of our contemporary psychological, social, and ecological ills. Although Jeffers describes his ideas about inhumanism as being philosophical in content, his presentation of those ideas is resolutely poetic and artistic in form. His decision to opt for writing poetry over prose (by the latter term he intends to denote not all non-poetic writing but rather the idle chatter of everyday discourse) is deliberate and strategic (Jeffers 1988–2001, 4, 391). He finds in poetry the kinds of movement and structure that emulate the rhythms (bodily, tidal, seasonal, planetary, cosmic, and so on) of the inhuman reality and beauty toward which he believes we should turn (Jeffers 2009–2015, 1, 685). There is also in poetic language, he suggests, a way of remaining on the side of inhuman things through

[2] Such a posture does not exclude, of course, full recognition of the existence of such outsides in other temporal registers.

marking the limits of human language and the very fact of human thought itself (Jeffers 1988–2001, 2, 409). Thus, Jeffers's poetry not only calls for a turning outward of our loyalties and loves, it enacts this turn in the very form of its verse.

In line with Jeffers's poetic and philosophical vision, the inhumanist approach I describe here takes as its goal a fundamental conversion in our individual and collective subjective dispositions and aesthetic sensibilities. By *aesthetic*, I have in mind the common use of the term as denoting the domains of beauty and the arts but also the etymological sense of *aisthēsis* as naming the general capacity for perception and discernment. In this sense, inhumanism is ultimately a matter of shifting one's attention and perception—and, ultimately, one's loyalties and loves—to the beauties and majesties of more-than-human registers of existence, which is to say, precisely to those ontological zones that have been delimited by the status quo as sacrificeable. Through inhumanist *aisthēsis*, the more-than-human world is glimpsed anew, viewed now as constituting a realm that precedes and exceeds the orbit of the human and its concerns. With inhumanism, the human (*anthrōpos*) no longer serves, contra Protagoras, as the measure (*metron*) of all things (*pantōn chrēmatōn*); instead, the more-than-human world stands as the measure and referential context of significance whereby another sense of the human emerges. Inhumanist *aisthēsis* reveals that the beings and practices associated with the human are but a partial and contingent mapping of an incredibly small slice of social, planetary, and cosmic existence. This approach thereby aims to re-inscribe the hegemonic anthropocentric order within a larger frame, wherein that dominant order is ultimately seen as indistinct from the sacrificeable realm from which the human seeks to distinguish itself.

Inhumanism offers a stark contrast to liberal humanism and its strategy of seeking to "raise up" animals into the subject position of the human. Inhumanism suggests that the path beyond sacrificeability is to be found in contesting the very notion that there *is* a sharp ontological or normative break between human and animal (or, for that matter, between the human and its variety of other "others"). For the inhumanist, the liberal humanist strategy employed by some pro-animal discourses which proposes that (certain) animals can be aligned with human beings while leaving other (less human-like) animals and non-animal entities in a sacrificeable realm is fundamentally objectionable. Indeed, the "good news" of inhumanism is that there are no inherently sacrificeable beings—that the very idea of creating a zone in which beings of any sort are to be seen as merely sacrificeable to an exclusively human order (or even to an order that has been expanded to include sentient animals) is precisely the onto-normative gesture that must be avoided. From the inhumanist perspective, the chief failure of the liberal humanist approach to animal justice is that it merely shifts the boundaries of sacrificeability and thereby ultimately remains committed to a sacrificial logic of establishing zones of inclusion and exclusion.

Inhumanism departs from zoo-pessimism as well in emphasizing the limits of the established anthropocentric order in terms of the presence and circulation of meaning and values and the constitution of worlds and relations. To be sure, an inhumanist approach has no interest in denying the substantial presence and influence of anthropocentrism (a point that the zoo-pessimist is right to emphasize); yet, acknowledging the force of anthropocentrism need not lead to overlooking the fact that the world is far richer in meaning, relation, and value than the anthropocentric order can fathom. Recognizing and affirming this more-than-human ex-orbitance is precisely what the inhumanist believes is necessary for moving beyond the logic of sacrifice and adopting alternative sensibilities, postures, and modes of perception.

The Art of Pulling the Emergency Brake

We can catch sight of key elements of this inhumanist sensibility in the remarkable expansion of interest in, and production of, roadkill art over the past two decades.[3] Roadkill artists such as Bobby Neel Adams, Marian Drew, Viivi Häkkinen, Marcel Huijser, Joy Hunsberger, Emma Kisiel, L.A. Watson, and Kimberly Witham (to name just a handful of the artists who have inspired my reflections in what follows) have created artworks involving roadkill that attempt to draw attention to the forces—both social and subjective—that lead to the notion of animals as mere obstructers of traffic. Their artworks strive to present roadkilled animals to the viewer in ways that standard habits of modern systems of mobility preclude (Watson 2015). Roadkill that might, at most, be merely glimpsed as one speeds by on a country highway or suburban freeway is presented in these works in such a manner as to recall that animal's life and death, its

[3] My analysis of roadkill art will necessarily be selective. For a broader overview of these developments, see Desmond (2016, 148–155). Relevant examples of roadkill art can be found readily online by searching for the work of the artists I mention in this section.

vitality and mortality, its mangled limbs as much as its uncanny beauty. By remaining behind with and recalling us to the animals who have been struck and left for dead on our roadways, these artists effectively serve as agents of Walter Benjamin's "emergency brake," seeking to slow and even jam the ever-more-rapid machine of hypermobility and redirect viewers onto another historical path (Benjamin 2003, 402).

These practitioners acknowledge that photographing, displaying, and re-presenting the bodies of roadkilled animals is a controversial practice, especially given the sometimes callous use of animal bodies and materials by some contemporary artists (Aloi 2012; McHugh 2011; Watson 2015). Yet, the bulk of the artists I have mentioned above insist that their artistic employment of roadkill is intended to demonstrate and enact deep respect for the animals depicted, not to denigrate or make light of their fate. Thus, Bobby Neel Adams describes the aim of his work as "paying homage to the creatures we are obliterating" through our driving practices (Minton 2018). Similarly, Kimberly Witham acknowledges that her work displaying dead animals walks a fine line of acceptability and might be seen by some viewers as offensive. She emphasizes, though, that "showing reverence for the creatures pictured" is of utmost importance for her work (Ching and Ching 2017). Such sentiments of veneration, love, and gratitude are expressed repeatedly by these roadkill artists and underscore their profound desire to render animals something other than, something more than, beings who are merely lose-able. By paying homage to the deaths of animals in their work, roadkill artists thus provide a performative enactment of a different kind of relation with the lives and deaths of animals.

At first glance, it might be tempting to interpret much of contemporary roadkill art as an instance of the liberal humanist sentiments we mentioned above. From this perspective, it might appear that roadkill art is an attempt to say that animal lives should be seen as being akin to human lives and as deserving the same respect and dignity. To be sure, something of this humanist sentiment is operative here. Nevertheless, many of these artists insist that what is at stake in their work is recalling human beings to their mortality and finitude, conditions that exceed the orbit of the human and that are shared with other animals (Häkkinen 2018; Kisiel 2014; Watson 2015; see also Marian Drew's remarks in Rosenberg 2013). They believe that we ignore the deaths of roadkilled animals in much the same way that many of us try to hold our own vulnerability, fragility, and death at a distance. In effect, then, roadkill art of this sort is not asking us to affirm that human beings have a unique relation to death or that some animals might share a "human" understanding of death (which is the temptation of some liberal humanist animal advocates). Rather, this art endeavors to displace us from our all-too-human perspectives on the carnage and death caused by our systems of mobility and to encourage us instead to inhabit a shared, indistinct, inhuman space wherein mortality and finitude circulate indiscriminately and without regard to any supposed human distinctives.

The viewer might be tempted, conversely, to see a distinctly melancholic affect running through much of contemporary roadkill art. There is a certain inevitability and intractability about the problem of roadkill that the artists seem to wish to bring to the surface, a reminder to the viewer that these deaths are the inevitable consequence of contemporary hypermobility. None of the artists I have mentioned present their work, after all, as being capable of *solving* the problem of roadkill, only drawing our attention to it. In this way, roadkill art might be read as being fundamentally pessimistic in nature and intent, an act by the individual artist of paying one's respects to dead animals, while lamenting the inevitable violence of the established order. Such art aims not at changing reality, it might be assumed, but rather serves as pessimistic confirmation of the ongoing violence and terror of anthropocentrism.

We should not dismiss too hastily the possibility that some roadkill art functions in precisely this pessimistic register. At the same time, the artists I have referred to here emphasize that their work is ultimately about encouraging viewers to attend to and affirm the transformative *beauty* of animal life—and death. In re-framing roadkilled animals as not just mattering but as manifesting beauty in both life and death, these artists recall us to the mundane but nevertheless astonishing fact of animal embodiment. They redirect our attention to the sheer fact *that animals are there* and thereby seek to convert our gaze from all-too-human affairs and toward the overlooked lives, deaths, sufferings, and beauty of the animals littering our roads (Hunsberger n.d.). Here, roadkill is temporarily displaced from a logic of sacrifice and seen to circulate, however briefly, in another register of meaning—one that is inhuman and more-than-human, thereby bearing witness to the fact that the dominant anthropocentric circuits of everyday meaning are not exhaustive or exclusive.

Again, there should be no naïveté about the salvific role of such art in addressing the problem of roadkill (Watson 2015). None of the artists mentioned here would maintain that art alone suffices for actualizing the sorts of fundamental transformations needed to address the scope of the problems surrounding contemporary hypermobility and the production of roadkill. But what *is* being suggested by this work is that something like a conversion in our individual and collective subjectivity and modes of *aisthēsis*—a conversion away from the logic of sacrifice that structures our average, everyday lives and toward inhuman and more-than-human reality, value, meaning, and beauty—is a necessary step on the path toward building a genuinely non-anthropocentric way of life. How widely such a way of life might be adopted, or whether such an aesthetic revolution at some scale is even possible, are not questions for the artist. Instead, they are tasks to be assumed and forms of life to be invented in response to the inexhaustible gift of animal life-death witnessed in the event of roadkill.

References

Aloi, Giovanni. 2012. *Art and Animals*. New York: I. B. Tauris.

Benjamin, Walter. 2003. Paralipomena to 'On the Concept of History'. In *Selected Writings*, ed. Howard Eiland and Michael W. Jennings, vol. 4, 401–411. Cambridge, MA: Harvard University Press.

Butler, Judith. 2009. *Frames of War: When Is Life Grievable?* Brooklyn, NY: Verso.

Cavalieri, Paola. 2001. *The Animal Question: Why Nonhuman Animals Deserve Human Rights*. Translated by Catherine Woollard. New York: Oxford University Press.

Ching, Darren, and Debra Klomp Ching. 2017. Interview with Kimberly Witham. http://atlengthmag.com/photography/kimberly-witham. Accessed March 1, 2022.

Derrida, Jacques, and Jean-Luc Nancy. 1991. 'Eating Well,' or the Calculation of the Subject: An Interview with Jacques Derrida. In *Who Comes After the Subject?* ed. Eduardo Cadava, Peter Connor, and Jean-Luc Nancy, 96–119. New York: Routledge.

Desmond, Jane. 2016. *Displaying Death and Animating Life: Human-Animal Relations in Art, Science, and Everyday Life*. Chicago: University of Chicago Press.

Franklin, Sarah. 2017. Staying with the Manifesto: An Interview with Donna Haraway. *Theory, Culture & Society* 34 (4): 49–63.

Häkkinen, Viivi. 2018. Artist's Statement from 'Forget Me Not'. http://muybridgeshorse.com/2018/08/09/viivi-hakkinen. Accessed March 1, 2022.

Haraway, Donna J. 2008. *When Species Meet*. Minneapolis: University of Minnesota Press.

Hill, Jacob E., Travis L. DeVault, and Jerrold L. Belant. 2019. Cause-Specific Mortality of the World's Terrestrial Vertebrates. *Global Ecology and Biogeography* 28 (5): 680–689.

Horkheimer, Max. 2004. *Eclipse of Reason.* New York: Continuum.

Hunsberger, Joy. n.d. Roadkill Manifesto. http://joyh.com/PHOTO/ROADKILL/body_roadkill.html. Accessed March 1, 2022.

In Defense of Animals. n.d. Wildlife Crossings—How They Protect Individuals, Species, and Ecosystems. https://www.idausa.org/campaign/wild-animals-and-habitats/wildlife-crossing/. Accessed March 1, 2022.

Jeffers, Robinson. 1988–2001. *The Collected Poetry of Robinson Jeffers*, 5 vols. Edited by Tim Hunt. Stanford, CA: Stanford University Press.

———. 2009–2015. *The Collected Letters of Robinson Jeffers*, 3 vols. Edited by James Karman. Stanford, CA: Stanford University Press.

Kisiel, Emma. 2014. At Rest. http://www.emmakisiel.com/at-rest. Accessed March 1, 2022.

McHugh, Susan. 2011. Stains, Drains, and Automobiles: A Conversation with Steve Baker About Norfolk Roadkill, Mainly. *Art and Research: A Journal of Ideas, Contexts, and Methods* 4 (1): 1–14.

Minton, Eric. 2018. Artful Decay: Bobby Neel Adams' Photography in Mourning. https://www.ppa.com/ppmag/articles/artful-decay-bobby-neel-adams-photography-in-mourning. Accessed March 1, 2022.

People for the Ethical Treatment of Animals. n.d. How Can I Help Prevent 'Roadkill'? https://www.peta.org/about-peta/faq/how-can-i-help-prevent-roadkill/. Accessed March 1, 2022.

Redmalm, David. 2015. Pet Grief: When Is Non-Human Life Grievable? *The Sociological Review* 63 (1): 19–35.

Rosenberg, David. 2013. Making Roadkill Beautiful. https://slate.com/culture/2013/02/marian-drew-photographing-still-lifes-of-dead-animals-in-australia-photos.html. Accessed March 1, 2022.

Schwartz, Amy L., Fraser M. Shilling, and Sarah E. Perkins. 2020. The Value of Monitoring Wildlife Roadkill. *European Journal of Wildlife Research* 66 (18). https://doi.org/10.1007/s10344-019-1357-4.

Seiler, Andreas, and J.-O. Helldin. 2006. Mortality in Wildlife Due to Transportation. In *The Ecology of Transportation: Managing Mobility for the Environment*, ed. John Davenport and Julia L. Davenport, 165–190. New York: Springer.

Soron, Dennis. 2008. Road Kill: Commodity Fetishism and Structural Violence. *TOPIA: Canadian Journal of Cultural Studies* 18 (1): 107–125.

Stanescu, James. 2012. Species Trouble: Judith Butler, Mourning, and the Precarious Lives of Animals. *Hypatia* 27 (3): 567–582.

Watson, L.A. 2015. Remains to Be Seen: Photographing 'Road Kill' and *The Roadside Memorial Project*. In *Economies of Death: Economic Logics of Killable Life and Grievable Death*, ed. Patricia J. Lopez and Kathryn A. Gillespie, 137–159. New York: Routledge.

Williams, Jeffrey J. 2010. Science Stories: An Interview with Donna J. Haraway. *The Minnesota Review* 73–74: 133–163.

CHAPTER 4

Subjects of Roadkill

Abstract This chapter focuses on Barry Lopez's essay "Apologia" in order to consider more carefully the issue of subjective formation and re-formation. In this essay, Lopez describes a cross-country road trip in which he encounters and removes a large number of roadkilled animals. Lopez describes this practice of roadkill removal as a "technique of awareness," a technique which I argue allows not just for fuller appreciation of the beauty and singularity of animals, but also serves as a means whereby a change in hyperautomobile subjectivity might occur. As a means of articulating this latter idea more fully, Lopez's ideas about apologizing and making amends are brought into conversation with related notions concerning events, truth, and ritual in Alain Badiou, Michel Foucault, and Emmanuel Levinas.

Keywords Apology • Ritual • Event • Subjectivity

My aim in this chapter is to pause for a moment before getting back up to speed and to bear in mind the interruptive hiatus to hyperautomobility provided by the inhumanist aesthetics of roadkill art discussed in Chap. 3. Thus, rather than accelerating directly toward a discussion of practical solutions to the problem of roadkill and other forms of mobility injustice (a task that will be taken up at more length in Chap. 5), I want to pursue

M. Calarco, *Reflections on Roadkill between Mobility Studies and Animal Studies*, https://doi.org/10.1007/978-3-031-30578-8_4

in this chapter a close reading of Barry Lopez's essay "Apologia" (Lopez 1998), which focuses on a number of his encounters with roadkill on an extended road trip. Examining this essay might seem to constitute something more of a digression or reversal in the course of my argument than a pause, for not only does Lopez's essay recount his cross-country car trip (such excessive driving seems to be the very problem at issue discussed in the present work), but he is himself responsible for some of the roadkill he examines. Further, Lopez's essay studiously avoids any discussion of how to solve the problem of roadkill at the socio-political level. No tactics are discussed; no short- or long-term strategies are mentioned; no policy initiatives or technological interventions are analyzed. Rather, Lopez simply narrates his road trip and the encounters he has with roadkill along the way.

What unfolds over the course of Lopez's essay, though, is a subtle consideration of a matter that is crucial for determining how and what sorts of experiments with altermobilities are undertaken: namely, the issue of the disposition and posture we adopt in addressing the problem of roadkill. Lopez's consideration of the issue of our subjective disposition allows us to raise in turn a number of important questions: Is roadkill a problem that can be fixed using primarily policy and technology initiatives and concerning which we can then wash our hands? Might reflecting on roadkill allow us to glimpse something of a deeper problem with our very selves, a problem with which many of us have deliberately avoided engaging? Is it possible genuinely to address the problem of roadkill without also reflecting on the subjective habits and dispositions that gave rise in part to the problem in the first place? Further, can subjects who have been thoroughly formed by the habits of hyperautomobility sustain such transformations without changing those habits and developing a taste and a passion for other ways of moving-through and being-in the world? It is toward such questions, I suggest, that Lopez's essay directs us.

Making Amends

Lopez's essay opens with him leaving his home in the Cascades in Oregon and heading east on a cross-country trip toward his destination in Indiana. Just a few miles underway, he encounters the first two of the many road-killed animals he sees on this trip. They are raccoons, and he carries them to the side of the road and lays them in the grass in a barrow pit. A bit further on and still in Oregon, he spots a series of five black-tailed jackrabbits and a young porcupine. He stops his car and carries each dead animal into the grass or brush at the side of the road. Lopez notes he is not entirely sure why he engages in this practice of roadkill removal—perhaps out of a sense of "decency" and "worry" (Lopez 1998, 113). He is also not entirely sure what roadkill *is*. As he carries the animals to the side of the road, he wonders: "Who are these animals, their lights gone out?" (113). To this, we could add the questions and concerns we have been developing in the previous chapters: How best to think about these animals? How best to relate to them? And can we even begin to answer these questions without our having some sense of the relations and worlds to which they belonged before they were struck? Along these lines, Lopez asks himself: "What journeys have fallen apart here?" (113).

Lopez remarks that he does not carry out this practice of roadkill removal consistently. After all, to do so would be time-consuming (especially in rural areas where there is considerable wildlife) and possibly risky (depending on the traffic and road conditions). What is more, the grief that is engendered through this practice can be overwhelming. Lopez notes that he tends to wince and spontaneously look away from roadkill when he catches a glimpse of it, preferring to imagine the vital life energy still pulsating through the animal's body. Once a roadkilled animal is in his hands, he is struck by how quickly that life energy fades from the body in motion but how long it remains in trace form among the eyes and fur of an animal. That roadkill is ontologically ambiguous is not something that will be lost on those readers who share in some version of Lopez's practice. Roadkill, as I argued in Chap. 2, is extraordinarily difficult to categorize due to its liminality. When dealing with freshly killed animals in particular, one often feels lingering traces of life, and yet the body seems rapidly to become something more like an inanimate object. While removing raccoons and a fox to the side of the road, Lopez notes that they "carry like sacks of wet gravel and sand" (114); in dragging a heavy doe off the road by its ears, he characterizes its body as a "gunnysack of plaster mud" (115).

In reflecting further on why he engages in the practice of roadkill removal, Lopez recalls that a man once asked him: "Why do you bother?" Lopez responded coyly to him, saying, "You never know.... The ones you give some semblance of burial, to whom you offer an apology, may have been like seers in a parallel culture," and then added, more straightforwardly: "It is an act of respect, a technique of awareness" (114). Here Lopez seems to be suggesting that by giving the dead animal a burial and an apology, one enters into an economy of the gift with that animal, and perhaps something of their abilities or insights as a seer will be given to you, but Lopez seems only half-serious in offering this response and does nothing to explain or explore it. The notion that he carries out this practice as an act of respect and technique of awareness would appear to be closer to one of the chief themes of the essay. But if we accept that the latter answer predominates, the question then becomes: What or who is being respected here? We might further inquire: Concerning what or who, precisely, does this technique make us aware? And what are the intended effects of such techniques on our own selves?

One could also wonder whether such a gesture is adequate for responding to the carnage created by our roadways. In Idaho, Lopez himself strikes an animal with his car, a young sage sparrow, and goes back to find its body. He checks the eyes and feels the body for life, but it has already passed. Undertaking his usual ritual, he lays the bird in a clump of grass on the side of the road. He nods before leaving, a gesture he makes out of grief but that leaves him feeling disconcerted, as if these acts of respect fail to rise to the gravity of the situation. In encountering two nighthawks that have been struck and killed while swooping down to catch gnats, Lopez stops to pick them up and move them from the roadway. He describes himself here as a "penitent" (114) while holding the two birds in his hands, despite the fact that he has not himself killed them. Rather than feeling personal culpability for these birds and most of the animals he encounters on the roadways, Lopez appears to be plagued by a sense of shame for his complicity with our established system of automobility. To be sure, he didn't kill the nighthawks himself, yet he might have under different circumstances; moreover, he knows he will undoubtedly kill other animals and life forms as he continues his drive to visit his friend. He may not be guilty of this particular killing, but he shares in our collective responsibility for participating in systems of automobility that are functionally structured by mass violence and killing.

There is something else at issue in Lopez's ritual practice of removing roadkill that is worth pausing over, namely, the question of animal beauty broached in Chap. 3. In attending to the previously mentioned nighthawks, which he describes as "lying soft as clouds in the road" (114), Lopez sees a farmer watching him at a distance and fears that the farmer will throw the birds back into the road after he is gone. Once again, Lopez seems to be concerned with roadkill sliding into the realm of "mere" objects—and not simply because they were once living beings. For Lopez, the animals he encounters are sacred—reminders of the unbridled beauty and remarkable nature of life. To Lopez's surprise, the farmer who is watching him from a distance does not want to throw the birds back but wants to share in the beauty of the nighthawks (which the farmer mistakenly thinks are actually whippoorwills, suggesting that beauty complicates and perhaps transcends taxonomical categorization in certain ways). Lopez notes that the farmer's voice is "weighted with awe," and that he "marvels" as he handles and inspects one bird's belly, bill, and wings (115). The farmer asks Lopez for permission to take the bird back to show his wife, hoping to share his fascination with her. He leaves the bird with the farmer, hoping that once the sense of wonder passes the bird will not end up in the trash, treated like a "whirligig" (115).

Ritualized handling and removal of roadkill also serves to reveal the singularity of animals. Lopez reflects on the singularity of roadkilled animals in several ways, but in two particular instances the unique nature of the specific animals he encounters strikes him with interruptive force. In the first instance, while driving through Wyoming Lopez hits a bird, despite swerving dangerously to try to avoid it. When he goes back to check on the animal, he is unable to identify its species with any precision. He sees that it is a gull, but turning it over in his hands, he can't recall the precise markings that would allow him to slot the bird into traditional classificatory schemas. But what he *is* able to discern is that this particular bird, which would seem to belong to an indistinct class of gulls, is there, dead, in his hands. He has no doubt about the damage that has been done to this specific body, "no doubt about the vertebrae shattered beneath the seamless white of its ropy neck" (116). Further down the road in Nebraska, Lopez spots a dead badger on the macadam. As he studies it, he notices its singular beauty, "the perfect set of its teeth in the broken jaw, the ramulose shading of its fur" (116). While such morphology and markings would normally function to help him place the animal in its proper biological category, in this instance he notes how the badger's markings differ slightly "as does every badger's" (116) from the kinds of drawings one sees in field guides. Thus, even as his naturalist background helps Lopez to appreciate the rich lives and relations of the animals he removes from the road, there is a way in which such knowledge is ultimately insufficient for doing justice to the singular life of a given animal. This gull, this badger, are not simply representatives of their species, but are unique beings situated in a distinct set of relations in a particular place. To sit with roadkill, to notice it, and to remove it respectfully from the road is one of the ways in which one might begin to recall oneself to the event-al nature of such specificities.

Lopez's roadkill removal ritual has the further, perhaps unintended consequence of bringing him into a different relationship with other drivers. While sitting with the badger, another driver in an old Chevrolet station wagon approaches Lopez from the opposing direction. The driver slows while passing and notices Lopez attending to the animal. As with the farmer, the driver shows no hostility toward Lopez but instead indicates his gratitude and shared grief. Lopez reads these dispositions from the bodily gestures of the driver as he passes: his hand "opens in a kind of shrug, hangs briefly in limp sadness, then extends itself in supplication. Gone past it curls into itself against the car door and is still" (116). Lopez takes this gesture to mean that he is not entirely alone in his grief for roadkill or in his unusual ritual practices; there do seem to be other drivers who care.

Yet, as his trip continues, Lopez becomes increasingly angry at his fellow drivers. In Nebraska, while picking up small mice and birds in the road, he refuses to "meet the eyes" of other drivers, for fear of the anger he will exhibit toward them. Lopez's anger might seem misplaced here, given that he was himself just responsible for striking and killing a gull. But the anger in this instance is more profound, and it is not clear that it is directed only at other drivers. To meet the eyes of other drivers and express his anger would undoubtedly lead to that anger bending back toward himself. Lopez knows that he is just as guilty as other drivers of treating animals' lives as if they were mere collateral, as if they were without value or meaning. In reflecting upon his anger, Lopez analogizes our collective acceptance of the killing of animals on our roadways with the killing of soldiers in war: we see both kinds of killing as "horrifying, unavoidable, justified" (116). And our acceptance of this sort of collateral damage leaves us "fractious, embarrassed" (116)—but only momentarily so. Inevitably, we return to our average, everyday lives and get underway again. Shortly after the stop in Nebraska, Lopez finds himself back on the road, where he strikes and kills a barn swallow. He finds it hanging "by its head, motionless in the slats of the grille" (116).

Pursuing further the thought of the sacrificeability of animals, Lopez stops in Nebraska to attend to a dead rabbit and notices a garter snake, an animal he might have otherwise missed had he not stopped. "What else have I missed," he asks himself, "too small, too narrow?" (116) Many of us spend our time on long drives observing large-scale things: mountains, landscapes, rivers, and so on. We do not typically study the roadways or roadsides looking for small creatures that we might hit or that are already dead; the design of the automobile and the structure of the roadways do not encourage that kind of attentiveness. It is worth emphasizing here again that cars are not something we simply drive and use as we wish, neutral forms of technology that can be used for "good" or "bad" and that allow our subjectivity to remain intact. Rather, they (and the systems within which they function) shape us in profound ways—forming our desires and attention at the most basic levels (Sheller 2004). As a practice, driving cues our field of vision almost entirely to the larger items in our purview and renders us functionally inattentive to that which cannot be seen at slower speeds and smaller scales. Lopez knows there are people who occasionally stop to help larger, more visible animals who are strewn across the roads. But are there others who have learned to slow down, move differently, and even pause to remain with the smaller, less visible creatures that occupy our roadways? And what of the insects who bear the largest brunt of the violence of our systems of mobility? Many drivers will be familiar with the uneasy feeling that arises when seeing the sheer number of dead insects that accumulate on one's car during a long road trip. At a motel stop in Iowa, Lopez scrapes from his grille numerous dead insects—bumblebees, wasps, butterflies—that have accrued on his trip. He does so, he says, because he is uneasy "carrying so many of the dead. The carnage is so obvious" (117).

Once his trip is over and his car finally comes to a rest in his friend's driveway, Lopez has a moment to pause and reflect on his journey as a whole. Sitting in the quiet of his parked car with the windows rolled down, a different series of connections with the world opens up. Lopez can now hear the insects outside his car, catching the loud song of the cicadas in an elm that he would have missed had he still been en route. Lopez is no doubt relieved to stop driving for a while, both in view of his temporarily not being directly responsible for any more killing and for not having to be witness to so much death on the roads.

There are aspects to driving, though, that go beyond the death and killing Lopez details. Long road trips through rural areas often expose us to some of the most memorable and beautiful landscapes and terrain we ever encounter, and Lopez does not deny the force of these experiences. At the end of his road trip, he recalls the "imposition of the Wind River Range in a hard, blue sky beneath white ranks of buttonhook clouds, windy hay fields on the Snake River Plain," and he writes movingly of the "transformation of the heart such beauty engenders" (117). But at what cost is such transformation created? Lopez contrasts the remarkable beauty of his drive with the "heavier memory" of the countless dead animals and insects he has encountered: "a catalog too morbid to write out, too vivid to ignore" (117). Although Lopez does not linger on this point concerning the beauty involved in and the joys of driving, it is one worth considering at more length. For if standard practices of automobility are to be transformed, such change cannot be simply a matter of reducing or stopping driving. It must be forthrightly acknowledged that driving is for many people (at times) a joyful and transformative activity, and for alternative modes of moving to take hold and become widespread, they must produce joys that are equally sweet and transformative. It will no doubt be necessary to undergo processes of resubjectification to learn to appreciate the beauty, joy, and meaning that attend other ways of moving, but such changes *can* be accomplished—especially if they are driven by a passion for more respectful relations.

But I am getting ahead of myself here, for consideration of these kinds of changes in our practices are held strictly outside the bounds of Lopez's text. The exclusion of such an analysis or any call for alternative modes of mobility might be frustrating for those of Lopez's readers who expect something of a manifesto against driving after his reflections on the carnage it causes. But Lopez's text is extremely subtle in terms of how questions are raised and where answers are provided or withheld. His text closes not with an examination of alternatives to driving (a topic I will take up in the following chapter) but rather with a set of strikingly spiritual reflections on what is at stake in his practice of removing roadkill. While standing at the driver's side of his car, and still in his friend's driveway before going in to meet him, Lopez experiences a sort of split in his subjectivity. He grips the window sill and leans down as if to talk to someone still sitting there—perhaps himself, perhaps another driver, perhaps all drivers. He is searching, he says, for words of atonement, searching for some way to alleviate the burden he feels but cannot fully articulate. For, it is not simply guilt or shame over the deaths of animals on the roadways from which Lopez desires relief; it is something much more basic and profound. At issue is a deeper grief concerning the ways of the world, "a sorrow over the world's dark hunger" (118). But no words of atonement, whether from a priest or from one acting as one's own priest, can lift this weight. Such sorrow, and the relentless realities that give rise to it, cannot be simply spoken away.

As Lopez's friend comes outside to welcome him in to his house, Lopez's reflections are threatened with interruption. He is left to ponder the question: "What is to be done with the desire for exculpation?" (118) His friend remarks that Lopez is a bit later in arriving than he thought he would be. Before responding to his friend, Lopez ventures an answer to his question about exculpation. He silently remarks to himself: "I do not want the lavabo. I want to make amends" (118). He then coyly tells his friend that his late arrival is due to his making more stops than he had planned to make. The essay closes with him leaving his car behind to join his friend inside for a chat. Lopez anticipates that conversing with his friend will serve as an "antidote," a kind of remedy or a cure for his sorrow. By bringing Lopez back within the "forgiving embrace of the rational" (118), the *logos* of conversation (as opposed to making amends) will serve to re-establish the familiar order of things, re-constitute the traditional structure of average, everyday life, and reassure him of "the human enterprise" (118). This reversion back to the quotidian is, to be sure, the constant temptation that pursues Lopez throughout this piece and that pursues us, his readers, as we move through the world.

Apology, or Becoming Worthy of the Event

We might read Lopez's reflections here less as offering an articulated alternative to dominant practices of automobility and more as developing considerations on what it might take to prepare oneself for committing to genuine transformations in our relation to driving, roadkill, and the "world's great hunger" that give rise to this form of mass killing. In other words, we should resist the temptation to read Lopez's text quickly or impatiently, looking for a thesis or a clue as to what alternative practices of mobility we should adopt. Although it is important to challenge dominant ideas concerning mobility and transform our practices accordingly, there is a danger in moving unreflectively from awareness of the problem to adopting alternatives (an issue that I will broach from another angle in my analysis of Giorgio Agamben in Chap. 5). Thus, rather than offering us a solution to the issue of roadkill that would make sense within the frame of our current mode of subjectivity, Lopez's text places himself and the reader in a zone of (im)potentiality and decision and poses a fundamental question: Will we seek to reanimate our lives in view of making amends for our everyday way of life, or will we return to the reassurance and comfort of "the human enterprise"? It is perhaps only by assuming the former path that we can undergo a genuine process of resubjectification, and it is perhaps along this path alone that a genuine encounter with the event of roadkill becomes possible. Stated otherwise, if our attention to roadkill is animated by the hope of making amends, the possibility arises for seeing roadkill as an event—which is to say, as an apocalypse (in the sense of an unveiling) of the underside of our own world, as well as the revealing of an unseen series of worlds and potentialities to which we are called to respond and negotiate. If, by contrast, we encounter roadkill strictly against the backdrop of the dominant form of anthropocentric mobilization, we will either miss its significance entirely or we will try to remove it from sight so that we can maintain our present way of life. It is these latter temptations against which Lopez is struggling throughout his work.

The suggestion I am making—namely, that the same roadkill will be encountered differently depending on the subjective disposition animating the encounter—will no doubt seem strange to some readers. After all, the roadkilled animal we see before us is a mind-independent entity, a brute fact, that precedes and exceeds my encounter with it. To suggest that there might be different "truths" about roadkill would seem to imply a sort of epistemological relativism. But the "truth" of roadkill at issue here is not so much an epistemological matter of gaining access to a series of facts about the animals we see on our roadways as it is a sort of "existential" truth. At stake is catching sight of and being true to the transformative force of an event, to beings and relations to which we are called to respond. In this sense, truth (as Alain Badiou has suggested) is a matter of living in such a way that we maintain fidelity to an event and allowing that event to work us over and transform us from within.[1] From a related perspective, Michel Foucault describes this sort of existential commitment as an instance of living a "true life" (Foucault 2005). In his final lecture courses (see, especially, Foucault 2010, 2011), Foucault becomes increasingly taken with the ancient philosophical idea that what is at issue in living a worthwhile life is precisely this enactment of and fidelity to processes of resubjectification, this following through on a spiritual *metanoia* that gradually and fundamentally transforms one's entire disposition (Hadot 1995). The sort of truth at issue here, then, is not something that is

[1] For a helpful overview of Badiou's complex ideas on this theme, see Badiou and Tarby (2013).

universally accessible to any inquiring subject no matter its everyday habits and mode of life. One has to undergo various forms of *askēsis*, modes of practice and training aimed at changing one's basic dispositions and redirecting one's "soul," in order to prepare for it and to remain committed to it.

One example in Lopez's text of an exercise of this sort is, of course, the technique of awareness involved in the ritual practice of removing animal bodies from the road. Now, this sort of roadkill removal practice (and related practices) can be carried out in two very different ways. On the one hand, the technique might be used to remove roadkill from sight and from thought. Once obligations for the dead have been met, one can then get back underway and move on with one's original projects. In this sense, the ritual of roadkill removal becomes an *apologia* in the etymological sense of the term, a defense speech, an attempt to "speak away" charges and to seek exculpation. With an *apologia* of this sort, there is no genuine affirmation of the encounter with roadkill, no attempt to be worthy of the event to which one is exposed. Rather, ritual here serves as a means to help us flee a sense of responsibility for what we are doing and for our way of life.

On the other hand, ritual techniques of awareness can serve as an apology in the more colloquial sense of the term, as a genuine expression of grief and sorrow, and as an attempt to make amends for what one has done and for the compromises that have been made in existing within the established order. Ritual here functions in the sense that Emmanuel Levinas grants the term in one of his early essays, "The Meaning of Ritual" (Levinas 2005 [1937]). In this text, rituals are understood to serve as a way of preventing us from stepping into the world unreflectively and uncritically, as a means of interrupting "for an instant the current that constantly connects us to things" (Levinas 2005, 288). Rituals, Levinas proposes, help us pause and consider our relations with others of all sorts (in this way, his early work casts ethical life in a much broader frame than the strictly and narrowly interhuman focus of his more mature writings that were touched on in Chap. 2); as such, they remind us of the sheer "miracle" of existence, of the fact "that the world is there" (288), that it is manifest to us, and that we are affected by it (to underscore a theme broached in Chap. 2 in relation to Heidegger). Driving and other hypermobile practices tend to function like the technological thread or current Levinas mentions, as circuits and connections that make us unreflective about and insensitive to the more subtle and fragile nature of the bodies and relations populating the landscapes that frame our ventures. To pause and sit with roadkill, to remove it from the road, to consider its life and death and relations, to assist the body's re-entry into trophic chains and new forms of life, death, and relations—these and similar rituals are precisely the sorts of practices that help to engender a subjective conversion and reconstitution, a genuine *metanoia* and rethinking of how one's subjective attention is currently directed and how it might be redirected. They serve as the concrete means whereby we are reformed, and it is through such ongoing, transformative practices that our habits and subjective dispositions are gradually realigned. Such practices are all the more important in contemporary contexts, where apparatuses of mobility have the potential to become all-encompassing and to prevent the space for critical reflection and the development of alternative ways of living and moving-in the world.

The subject that emerges from these latter kinds of ritualized practices has the potential, then, to be very different from the one who began them. The reformed subject is the subject of a truth, one who has been touched by the event of animals, and who has begun the ongoing task of apprehending the entire arc of the life and death of roadkill, the broader systems of power that override this arc, and the need for a genuine alternative to current systems of mobility. The reformed subject is now better prepared—better disposed, we might say—to consider what it might mean to move otherwise through the world and in ways that will not continue to cover over the violence of our systems of mobility but bear them habitually in mind. It should be stressed, though, that resubjectification of this sort carries no strict chronological priority; it is not as if one needs first to undergo and complete this sort of resubjectification in order to begin addressing the larger subjective, social, and environmental issues surrounding roadkill. Living a "true" life always begins *in media res* and with the tools and potentials one finds at hand. But even if there is no chronological priority concerning processes of resubjectification, there is a *logical* necessity at work here. For the sort of fidelity and militancy required both to grasp the revolutionary changes needed to address roadkill as well as the passion to sustain the struggle for such changes over the course of a lifetime is not something that can be achieved by those who remain predominantly subjects of hyperautomobility. The problem of roadkill demands of us that we become something other than who we are, born again, in response to an event that calls us to another way of life.

References

Badiou, Alain, and Fabien Tarby. 2013. *Philosophy and the Event*. Translated by Louise Burchill. Cambridge: Polity.

Foucault, Michel. 2005. *The Hermeneutics of the Subject: Lectures at the Collège de France*. Translated by Graham Burchell. New York: Palgrave Macmillan.

———. 2010. *The Government of the Self and Others: Lectures at the Collège de France, 1982–3*. Translated by Graham Burchell. New York: Palgrave Macmillan.

———. 2011. *The Courage of Truth (The Government of the Self and Others II): Lectures at the Collège de France, 1983–4*. Translated by Graham Burchell. New York: Palgrave Macmillan.

Hadot, Pierre. 1995. *Philosophy as a Way of Life: Spiritual Exercises from Socrates to Foucault*. Translated by Michael Chase. Malden, MA: Wiley-Blackwell.

Levinas, Emmanuel. 2005 [1937]. The Meaning of Religious Practice. *Modern Judaism* 25 (3): 285–289.

Lopez, Barry. 1998. Apologia. In *About this Life: Journeys on the Threshold of Memory*, 113–118. New York: Vintage.

Sheller, Mimi. 2004. Automotive Emotions: Feeling the Car. *Theory, Culture & Society* 21 (4–5): 221–242.

CHAPTER 5

Profaning the Streets

Abstract This chapter considers genuine alternatives to a hyperautomobile way of life. Such "altermobilities," I argue, are not mere replacements for cars and car use but constitute very different ways of being-in and moving-through the world, ways that are attentive to our human and more-than-human kin. Thus, altermobilities are aimed at enacting ways of moving and being that are truly common and shared (what I will call, following the work of Giorgio Agamben, *profane*) rather than exclusive and hierarchical (or *sacred*, in Agamben's sense of the word). Through an analysis of the ideal of autonomobility (developed by mobility theorists Noel Cass and Katharina Manderscheid), along with recent developments in the fields of road ecology, the environmental humanities, and disability studies, I sketch the elements of a genuinely non-anthropocentric approach to thinking about and practicing mobility.

Keywords Profanation • Apparatus • Autonomobility • Road ecology • Environment • Vulnerability

On a remote road just outside the small town of Amherst, Virginia, I notice an enormous snapping turtle near the side of the road ahead of me. I slow my bicycle to let the turtle cross and notice that it is struggling to

M. Calarco, *Reflections on Roadkill between Mobility Studies and Animal Studies*, https://doi.org/10.1007/978-3-031-30578-8_5

walk. Despite the low traffic and the excellent visibility, it appears the turtle has been struck by a car (probably deliberately, I deduce). The damage to the turtle is extensive. Its shell has been crushed on the left side, and its feet on that same side have been sliced open and are bleeding. Still, it is moving determinedly across the road, in the direction of the Tye River.

Fearing that it will be struck by another car, I decide to dismount and try to redirect the turtle off the road and into a ditch. Due to the damage to its shell, I can't lift it. Instead, I use a large stick and try to prod it gently in the direction I want it to go. The turtle snaps angrily at me and won't budge. It insists on crossing the road and continuing on its chosen path. After trying and failing for several minutes to get the turtle to stay off the road, I decide simply to sit down on the side of the road and let it go its own way. While waiting, I flag down approaching traffic and make sure the drivers see the turtle and go around. After several agonizing minutes, the turtle slowly shuffles down the bank on the opposite side of the road and continues on its path.

I am unsure whether that turtle survived; given the extent of its injuries, the prospect is unlikely. Sharing the story many years later with a naturalist friend from the area, she suggests that given the time of year, the turtle was likely trying to get to a specific spot to lay her eggs, hence the determination and unwillingness to yield to another path. We talked about the countless thousands of turtles of multiple species that cover the roads in the area that time of the year, how many are struck and injured or killed, and various strategies for limiting the carnage. I thought to myself that no amount of fencing or designated crossings (the standard mitigation strategies) could make a dent in those numbers. What the turtles need is to have their traditional pathways restored so that they can move at their slow but determined pace to get where they need to go.

My critical reflections on roadkill began in earnest around the time of this event. That was when I first began to piece together my fragmented thoughts about automobility on the one hand and animal life on the other and genuinely to think about roadkill and the harms that driving causes to animals and the more-than-human world. Living in rural Virginia at that time, I was witness to an inordinate amount of roadkill—more than I had ever witnessed growing up in the suburbs of southern California. Up to this point in time, I was certainly aware of roadkill, but I primarily wanted to remove it from sight and to assuage my guilt in relation to the problem; after this encounter with the turtle, I finally began to try to make amends for the role I was playing in causing this carnage and change my usual means of moving and driving. Shortly thereafter, I sold my car and began experimenting with alternative, and less destructive, modes of transport—an ongoing process that continues to the present.

* * *

In Chap. 1, I suggested that hyperautomobility and anthropocentrism are two modes of power that are essential for understanding contemporary forms of subjectification and analyzed how these two systems function to naturalize roadkill. In Chap. 2, I offered an alternative understanding of the ontology of roadkill, not as a substance or object, but as an event. In Chap. 3, I discussed the social and ontological dimensions of roadkill in terms of the logic of sacrifice and considered how the work of contemporary roadkill artists complicates that logic and re-directs our attention to the event of roadkill. In Chap. 4, I examined how processes of resubjectification are crucial for becoming the kinds of beings who desire to develop alternative ways of being-in and moving-in the world. The questions that remain to be addressed in this final chapter are: How best to contest and modify the specific systems and practices of mobility in which we find ourselves? And how to do so from a non-anthropocentric perspective? In taking up these questions, I turn first to an analysis of Giorgio Agamben's work on apparatuses and his corollary concept of profanation. I then consider the work of mobility theorists Noel Cass and Katharina Manderscheid, who provide a compelling vision of how to transform systems of mobility in the direction of what they call autonomobility, an ideal that I suggest is deeply consonant with Agamben's notion of profanation. Finally, I examine what it might mean to reframe these notions along non-anthropocentric lines and briefly consider some of the altermobilities to which such an approach might give rise.

Profaning Apparatuses

Agamben's essay "What Is an Apparatus?" commences with what appears at first glance to be a rather modest hypothesis regarding Foucault's influential concept of *dispositif*. Agamben proposes that *dispositif* (translated in the English version of the essay as *apparatus* and as *dispositivo* in Agamben's original Italian text) is a "decisive technical term in the strategy of Foucault's thought" (Agamben 2009, 1). According to Agamben, the concept of *dispositif* is neither neutral nor denotative; rather, it serves to bring a phenomenon or series of phenomena to presence, to wrest them from obscurity. Further, on Agamben's reading, this disclosive concept of *dispositif* functions productively for Foucault. A productive term or concept leaps ahead into a given field of forces and relations in order partially to delimit them and to open a field of meaning (Heidegger 1962, 30–31).

Agamben further suggests that, whereas Foucault's earlier key concept of an *episteme* was used to distinguish the scientific from the non-scientific within discursive systems alone, apparatuses deal with discourses of various sorts as well as non-discursive institutions, architectures, and so forth. More specifically, an apparatus refers to the *linkage* between various discursive and non-discursive elements that form a "system of relations that can be established between these elements" (Foucault 1980, 194). As with many of Foucault's critical terms from this period of his work, concepts like apparatus are meant to uncover dynamic and complicated forms of power relations. In particular, the concept of an apparatus aims to name a response by power to a problem, namely, the problem of governing the various behaviors or characteristics of a population, and in response to that need, various technologies and discourses are brought together in an ensemble, an apparatus, in order to establish control. Thus, we could say that apparatuses help both to delimit and regulate a given population, establishing relations of power that operate in both repressive and productive terms.

Of interest to both Agamben and Foucault here is not simply defining apparatuses but uncovering their basic *functioning*, which is to say, uncovering the ways in which they work to produce what appear to be "naturalized" subjects. As Agamben's notes: "The term 'apparatus' designates that in which, and through which, one realizes a pure activity of governance devoid of any foundation in being. This is the reason why apparatuses must always imply a process of subjectification, that is to say, they must produce their subject" (Agamben 2009, 11). The underlying process of subjectification that Agamben highlights here is something like the "essence" of apparatuses: they are a key means of governance and control; yet—and just as important—technologies of subjectification can also "accidentally" derail or lead to unanticipated challenges to governance, a point that will be of importance to my discussion in the second half of this chapter.

As a means of getting at the subjectification processes at work within apparatuses—and here Agamben moves beyond the modest hypothesis regarding Foucault that opens his essay and begins to develop his own distinctive approach—Agamben suggests starting from a broad ontological partition between *substances* (something like Aristotle's individuals, or primary substances) and *apparatuses.*[1] But what Agamben adds to this socio-political ontology is a third category, that of the *subject*, and it is this third category that allows him to disclose the inner logic of apparatuses. Thus, within this proposed ontological framework, there are (1) living beings (substances, or what Agamben sometimes calls "creatures," in a Benjaminian vein) on one side of the divide and (2) apparatuses on the other side, with (3) subjects emerging at the interstices of this matrix via the interactions and shapings that occur between living beings and apparatuses.

In terms of what Agamben means by "living beings" here, it is clear that he has a very broad scope of entities in mind, including and extending well beyond human beings—which constitutes one signal difference between his work and Foucault's inasmuch as Foucault is generally uninterested in tracking the play of power beyond human beings.[2] With regard to the scope of what constitutes an apparatus, here too Agamben differs from Foucault in allowing the concept of apparatuses to comprise a range of devices, practices, and institutions that is considerably broader than what Foucault tended to include under that label. Thus, Agamben can say that he would include under the heading of apparatus all of the well-known institutions and practices that Foucault examined (prisons, schools, medical institutions, and so on), assemblages that tend to function in the role of a network to bring together various registers of power, knowledge, and control; but he also understands apparatuses to include more basic technological devices that do not always play the organizing and network-type roles that Foucault tends to allot them. Under this broader rubric, Agamben includes "the pen, writing, literature, philosophy, agriculture, cigarettes, navigation, computers, cellular telephones, and—why not—language itself, which is perhaps the most ancient of apparatuses—one in

[1] Here we arrive back on the terrain with which we began our ontological analysis of road-kill in Chap. 2. As we will see, Agamben's critical reworking of the concept of substance allows for an understanding of both human and more-than-human beings who are caught up in apparatuses as constituting something more than traditional substances—that is, as beings characterized by an inherently dynamic and event-al structure, shot through with an (im) potentiality that renders them simultaneously close both to life and death.

[2] The status of animals within Foucault's larger oeuvre is a complicated subject I cannot broach here. For a fuller discussion of this topic, see the essays collected in Chrulew and Wadiwel (2017).

which thousands and thousands of years ago a primate inadvertently let himself be captured, probably without realizing the consequences that he was about to face" (14). That systems of mobility are prime examples of such apparatuses should be evident, and I will examine mobility from this angle in more detail below.

First, though, I want to focus more carefully on Agamben's third ontological category: the subject produced by apparatuses. In his analysis, a subject should be understood as that kind of being whose existence helps to sustain and reproduce a given apparatus; it exists as a being that is thrown-under (*sub-iectum*) an apparatus as its support. And given that most living beings encounter and are captured by multiple apparatuses during their lifetimes, they can be viewed as going through multiple processes of subjectification. This gap or difference between (a) the various modes of subjectification and (b) the substances that undergo subjectification is what requires Agamben to insist upon a third ontological category. What we might be tempted (following Aristotle) to identify as a singular substance—that is, a being that retains something of an identity across time—can, on Agamben's account, itself become multiple subjects. And this multiplicity does not make a given substance's varied subjectivities any less real—for the subjects and subject positions produced by apparatuses (Agamben gives examples of subjects such as "the user of cellular phones, the web surfer, the writer of stories, the tango aficionado, [and] the anti-globalization activist" [14–15]) have a material existence and constitute real effects. Indeed, substances often become deeply invested in these subjectivities, sometimes identifying and at other times dis-identifying with them; they also play an essential role in the organization of social life at multiple levels, including the psychic, social, and economic levels.

Social theorists in recent years have become particularly adept at showing how such processes of subjectification are, despite their ontological reality, contingent and open to being transformed. Subject positions and apparatuses, on the one hand, and the substances that are caught up within them, on the other, are not wholly aligned or strictly identical; there is something of an ontological fissure or fold between them. In order to understand what is at issue in Agamben's analysis here, it should be noted that he shares with contemporary social theorists a desire to uncover the contingency of processes of subjectification; but we have to be careful in how we figure this zone of contingency if we wish to follow his unique line of thought. In the context of Agamben's thought, it is important to emphasize the subtle point *that* there is a gap between being and action.

If we move too quickly past this *that*, this subtle fact of the fissure or fold, in order simply to substitute some new mode of subjectification to replace the old one, we risk losing sight of the deeply non-foundational nature of action and subjectification. In short, the fact *that* there is a gap between being and action, between substance and subject, indicates for Agamben that *there is nothing we must become* (a fact that cannot be recalled frequently enough in an age as restless and hypermobile as our own). This kind of slippage between substance and subject should not be understood (as the existentialist philosophers of the preceding generation argued) as a space of freedom but rather one of radical potentiality and im-potentiality. It recalls us to the simple but easily overlooked truth that living substances have neither a vocation nor a work (in the Aristotelian sense of an *ergon*) to which they properly belong, and it further indicates that any vocation, identity, or work which a living being takes up or with which it is identified is done so without a ground or foundation in its being. A subject emerges only in and through a confrontation and negotiation with the non-ground of inoperativity.

But what happens when the apparatuses with which substances interact become ever-more pervasive, granting ever fewer opportunities for glimpsing the gap between being and action? These are the concerns and questions that arise as we try to come to grips with the shift in contemporary societies toward new modes of governmentality and control. In modern capitalist societies, human beings are surrounded by and enveloped in apparatuses, effectively without break or rupture. As Deleuze notes, this is one of the characteristic features of control societies in contrast with disciplinary societies (Deleuze 1995). In the latter, discipline is certainly pervasive but there are at least occasional breaks and ruptures in the interstices between institutions; in the former, the apparatuses of power function in overlapping and seamless ways to create a functionally ubiquitous network of power relations that render reference to an "outside" effectively senseless. Agamben echoes this Deleuzean point when he writes: "It would probably not be wrong to define the extreme phase of capitalist development in which we live as a massive accumulation and proliferation of apparatuses. It is clear that ever since *Homo sapiens* first appeared, there have been apparatuses; but we could say that today there is not even a single instant in which the life of individuals is not modeled, contaminated, or controlled by some apparatus" (15). In such a situation, where individuals come increasingly to be integrated within networks of control, the possibility of glimpsing the slippage between being and action becomes slighter and slighter; concomitantly, the inoperativity and (im)potentiality of earthly existence goes increasingly unheeded, along with the possibilities for refusal and resubjectification that make other forms of life possible.

Agamben is, thus, raising the question of what form ethics and politics might take in an age that is increasingly characterized by networks and technologies of control. The two dominant critical responses to the ubiquity of control and technology—destruction of apparatuses on the one hand and their correct use on the other—strike Agamben as equally implausible. And although he admits to the temptation to destroy or eliminate certain technological devices such as cellphones (a temptation that indicates his distaste for apparatuses that homogenize social relations along reductive and one-dimensional lines), such an approach would overlook the facts that apparatuses are neither somehow unnatural or foreign to creaturely life nor are they utterly at odds with the happiness and flourishing of living beings. Recognition of these points ultimately leads Agamben to maintain that the proper response to the ubiquity of apparatuses in modern times is not destruction or correct use but *profanation*. Profanation is the process whereby apparatuses are removed from the sphere of sovereignty and control of the few (the realm of the *sacred*) and returned to immanent, common use and shared happiness (the realm of the *profane*). From Agamben's perspective, the chief problem with the apparatuses, devices, and institutions that crowd modern life is that they have largely emerged from and operate in the interest of a sphere considered to be separate from and transcendent to common social life. Properly reoriented to free and common use, apparatuses might become further denaturalized and thereby allow living beings partially to escape the nets of capture and control and once again to encounter their (im)potentiality. My task in the next portion of this chapter is to explore at more length this notion of profanation in regard to systems of mobility and to consider how it might be deployed both to refuse and reconfigure modern mobilities in view of both human and more-than-human flourishing.

Profanation as Autonomobility

Sociologists Noel Cass and Katharina Manderscheid's insightful critical analysis of contemporary systems of automobility provides an opening for understanding how the notion of profanation might be applied in this register. Under the rubric of *autonomobility*, they explore the ways in which hyperautomobility and the practices and subjective habits it inculcates in drivers both radically undercut human autonomy and marginalize large swaths of the human population. What is perhaps most interesting in Cass and Manderscheid's analysis is their joint attentiveness to the injustices of current systems of mobility coupled with a resolute refusal to accept standard liberal or pessimistic responses to those issues (variants of which I explored in relation to animal politics in Chap. 2). Thus, rather than resting content with endorsing minor reforms to dominant systems of mobility or throwing up their hands in the face of the ubiquity and inertia of these systems, they articulate what they call an "alternative imaginary" (Cass and Manderscheid 2019, 101) for future forms of automobility that is grounded in a variety of liberatory discourses and practices. Although I will ultimately seek to extend Cass and Manderscheid's imaginary in ways they might not countenance, I want to follow their analysis up to that point of departure, as I believe doing so will help both to develop the notion of profanation in the context of mobility systems in a useful way as well as to highlight the need to broaden the scope of profanation to the more-than-human world.

Similar to Agamben's critical approach to apparatuses, Cass and Manderscheid take a largely "dark" view of the power associated with systems of mobility. Contemporary forms of automobility are viewed by them primarily as repressive apparatuses of subjectification, and the processes whereby subjects are constituted are likewise seen as being largely exclusionary in nature. Thus, even if Cass and Manderscheid would be willing to grant the familiar Foucaultian point that apparatuses are both repressive and productive, their analysis is much closer in spirit to Agamben's in that the productive and liberatory dimensions of automobility remain largely to be created and wrested from those apparatuses rather than being understood as somehow latent in them.[3] Left to their own workings (or rather, left to the machinations of the capitalist economic order and the interests of the ruling class), systems of automobility have had, according to Cass and Manderscheid, a generally devastating effect on life in subjective, social, and ecological registers. They provide a litany of examples illustrating what they label the "dark sides" (104) of automobility, including its negative environmental consequences, high numbers of fatalities, and harms to health from air pollution. They also emphasize the ways in which marginalized peoples often

[3] These issues are more directly broached in Manderscheid (2014).

suffer compounded effects from these problems, given their already precarious modes of existence and common proximity to dense circuits of mobility and transport.

What is of primary interest for our purposes in Cass and Manderscheid's analysis is how their approach to addressing the problems with dominant mobility systems carves out a distinct space that differs from the standard paradigms in both transportation studies and mobility studies. Here, they largely reject the technofixes (such as electrification and automation) that are offered by mainstream advocates of transportation sustainability, solutions that effectively leave the more fundamental questions of infrastructure off the table. Along this line, one of the chief advances made by progressive advocates in the field of mobility studies is to probe this question of mobility infrastructure and to seek to render systems of mobility more equitable. This has given rise to a significant body of literature and activism around mobility justice, much of which is consonant in spirit with the concerns over the dark sides of mobility raised by Cass and Manderscheid.[4] Yet, for Cass and Manderscheid, simply making systems of mobility more equitable and accessible, while certainly an improvement over the reformism of technological approaches, is an inadequate strategy. What remains largely hidden from view in both the technologically oriented (moderate) and the justice-oriented (progressive) approaches to mobility reform is the question of mobility itself. In other words, both of these approaches take "the *need* for mobility as a fixed requirement of the good life without questioning the underlying compulsions to be (auto) mobile" (102).[5]

Those compulsions are brought more fully to the surface if we understand automobility as an apparatus in the sense described by Agamben—that is, as a system of technological and power relations that produce subjects who gain a sense of their being and meaning in and through their belonging to those apparatuses. In the process of becoming subjects of the car and of the rhythms of contemporary mobile life, the various compulsions to be hypermobile not only disappear from conscious awareness but begin to appear as if they are organic and natural outgrowths of

[4] Recently published works on mobility justice that are representative of these concerns include Martens (2016), Montegary and White (2015), Sheller (2018), and Vannini (2016).

[5] David Bissell's analysis of the phenomenon of waiting offers another angle from which to think about the subtle potentials and (im)potentials that trouble common compulsions to be mobile (Bissell 2007).

those subjects. For Cass and Manderscheid, contemporary capitalist economies play a manifest and primary role in effecting this subjectification, and they do so through the constitution and sustaining of both the bulk of the technological systems of automobility and the ideologies of freedom of movement and independence associated with a hypermobile way of life. Through naturalizing hypermobility, the compulsions that coerce individuals to assimilate to the system are taken for granted and hence go largely unnoticed and unchallenged. Thus, we accept such practices and conditions as long and time-consuming commutes; moving long distances for jobs; high numbers of fatalities and injuries associated with driving; and so on. Conversely, we often overlook the forced *im*mobility of those who are unable to access systems of mobility or afford the means to be hypermobile and fail to see the precarity of a life lived on the edges of or outside this system (Cresswell 2010).

I have already examined in previous chapters some of the micro-practices that might begin to break the grip of such subjectification; I will not retread that ground here. At stake in this chapter is considering the form that alternative mobilities and counter-systems might take if we wish to contest compulsory forms of mobility and immobility of the sort just mentioned. With this aim in mind, Cass and Manderscheid provocatively posit *autonomobility* as an ideal for mobility studies and activism rather than the more standard goals of accessibility, equity, or justice. With the concept of autonomobility, they mean to place the emphasis on autonomy and freedom in the negative sense of the term, that is, as being freed *from* compulsions of either mobility or immobility (and here, their work finds its natural complement in Agamben's concept of im-potential and his emphasis on contesting the commonplace assumption that there is something that human beings must be or do). Although seeking to maximize liberty, their framework is far from being libertarian in the individualist sense, for the ultimate goal of their approach is that of the mutual flourishing of individuals *and* collectives. Furthermore, they stress that autonomobility is not simply opposed to the ideals of accessibility, equity, or justice; rather, the point is that, if liberal ideals and reforms are not nested in an autonomobility framework, they will not genuinely address the compulsions that generate mobility-related injustices in the first place.

The counter-system of mobility that Cass and Manderscheid envision as growing from the premises and ideals of autonomobility is parasitic on and extends existing approaches to mobility championed primarily by radical environmentalists.[6] I will examine elements of this counter-system in more detail below, but before doing so, it should be noted that Cass and Manderscheid have no illusions about such an alternative being widely and readily adopted in the near future. They grant that the ideals and system they are promoting will appear utopian and unpragmatic when placed against the realities of present-day mobility systems. What their work attempts to do, though, is broaden our social imaginary by describing an alternative scenario and way of life that allows us to denaturalize our most basic assumptions about mobility. In other words, their aim is to expand and extend the future imaginary of radical environmental and sustainable modes of transport in such a way that autonomobility is allowed to infuse that vision with its own spirit.

In terms of concrete changes and planning, Cass and Manderscheid assume that the actualization of an autonomobility regime requires implementing the basic features associated with sustainable and localized forms of transport, such as the implementation of walking and cycling infrastructure, densification of urban areas, and the decentralization and localization of economic relations.[7] Such infrastructural changes would fundamentally lower the requirements for being hypermobile and therefore eliminate in advance the challenge of having to transform all existing circuits of transport. The development of these sorts of localized communities and economies also raises the thorny question of how best to deprivatize land ownership, for this slower, more localized way of life is driven not by growth and profit but by an extra-economic ideal of flourishing.[8] As the transition to localized economies is made, Cass and Manderscheid also anticipate the need for transitional modes of mobility of various sorts, as the infrastructure for truly local circuits of transport must in many cases be created over time and might be possible at first only for a handful of privileged and able-bodied people (Stehlin 2019). Over time, though, the goal would be a fundamental reduction in the very need for transport by decoupling the requirement to be mobile from economic stability, which would make possible slower and genuinely sustainable forms of transport.

[6] For a brief overview of this approach, see Dennis and Urry (2009, 149–151).

[7] This is a large field of research, but helpful points of entry include Banister (2005, 2008), and Tolley (2010).

[8] Here the anti-capitalist and quasi-anarchist leanings of Cass and Manderscheid simultaneously align and sit in tension with resurgent Indigenous critiques of land ownership. For fuller discussion of such points of overlap and difference, see Coulthard (2014) and Estes (2019). I will return briefly to the issue of land relations below.

Another distinctive theme in Cass and Manderscheid's autonomobility approach, and one that brings it into close proximity with Agamben's notion of profanation, is its emphasis on collectivizing transport. At the heart of the hyperautomobility regime is the single-occupant vehicle and its driver pursuing his own projects with little regard for aligning his own mobility needs with those of others. Collectivizing transport, by contrast, takes seriously the idea of roads and circuits of mobility as commons, not just in terms of being accessible to all, but also as a site where sociality might be renewed and reinvigorated. Profaning the streets in this collectivist sense would include displacing the dominance of the automobile in favor of reconfigured social spaces, a variety of modes of public transport, as well as vehicles and modes of mobility that allow for collective interaction. Examples of the latter would include everything from collective commutes (sometimes referred to as "bike buses" and "walking buses") to using bicycles and velomobiles that allow for increased interactions and conversations between users. Cass and Manderscheid also underscore the importance of legalizing and revalorizing various modes of sharing, hitch-hiking, and piggybacking, whether in the form of collective car shares, allowing people to free-ride on forms of transport that move goods, or increased numbers of group outings on buses and other multi-person vehicles. All of these alternative, collective modes of transport underscore the point that "'single-occupancy vehicles' are anathema to autonomobile society" (109).

That these and countless other examples of alternative, collective, and sustainable altermobilities are extant (Newman and Kenworthy 2015; Furness 2010) should indicate that we are not entirely in the realm of utopian theorizing here.[9] To be sure, no advanced industrialized states or cities (even the most progressive among them) have fully realized the sorts of alternative infrastructure and practices championed by Cass and Manderscheid. The point of their analysis, though, is not to argue that such alternatives can be immediately actualized on a large scale but rather to expand our future imaginary and to denaturalize the present system (as

[9] This fact should also be taken to indicate that social life—especially when the social bond is understood to extend beyond interhuman relations to encompass the more-than-human world—often contains more potentials than we are apt to assume. For a fuller discussion of such a stance—and one that is critical of the more pessimistic aspects of Agamben's work on apparatuses—see Stiegler (2010). As the reader will note, I am developing an argument related to Stiegler's but from a non-anthropocentric perspective that puts Agamben's terms to work in different directions.

suggested above). The latter task can take inspiration from the fact that these extant practices, however minor, are realistic pre-figurations of alter-mobilities, thereby providing seeds that can be sown to construct a path toward another way of life. At the same time, Cass and Manderscheid argue that autonomobile practices and ideals will always remain but a minor presence in our societies if the counter-force of capitalist ownership of systems of mobility remains in place. As with the single-occupant vehicle, capitalist norms regarding the movement of goods and people serve the interests only of a select (or "sacred," in Agamben's terms) class and, hence, are at odds with the genuine profanation of transport and mobility. The main obstacle standing in the way of the development of altermobilities, they maintain, is that dominant "social systems are currently organized for shareholders, not mobility requirements or other understandings of social need" (110–111).[10]

We might wonder, though, whether private ownership of mobility systems is the only or even central limit that must be addressed if the aim of our analysis is to profane the streets and adopt something like autonomobile ideals. For, despite the radical and thoroughgoing transformations that Cass and Manderscheid promote (many of which I would generally endorse), their autonomobility framework remains almost entirely anthropocentric and thus fails to address what it might mean to prioritize autonomy, new forms of sociality, and alternative modes of mobility among and in relation to more-than-human beings. Thus, even though they go well beyond many of the liberal and progressive dogmas of mainstream mobility studies, they share with this field its uncritical tendency to overlook the status of the more-than-human world, both in regard to analyzing the effects of mobility systems and proposing alternatives to those systems (Swart 2015). Cass and Manderscheid—like most mainstream mobility theorists—do underscore the environmental crises (climate change, pollution, extinction, and related crises) associated with hyperautomobility, but the environment is primarily understood in this discourse as the neutral and anonymous background against and through which human beings move. The more-than-human world has neither any standing of its own nor any agency or mobility needs of its own under this approach.

In order to address this limit, we might consider returning to Agamben's *dispositif* analysis, for his account of subjectivity and apparatuses is

[10] Further discussion of this point can be found in Scott (2020).

surprisingly capacious. Recall that, for Agamben, an apparatus is an assemblage of forces that governs living beings and which, through that governance, produces subjects of various sorts. That Agamben chooses the term "living beings" (rather than simply human beings) to describe the entities who are captured and formed by apparatuses no doubt speaks to his well-known interest in the way in which apparatuses capture *zōē* and mold it into the subjectified *bios* of human political subjects. But what if we took seriously the broader, more-than-human frame for understanding subjectification that Agamben provides and sought to consider the ways in which animals and other living beings (as well as inorganic beings, systems, and other more-than-human existents) are both captured by and resist apparatuses—and by apparatuses of mobility in particular?

If we were to do so, one of the first points that would come into view is that the stakes of profaning systems of mobility and apparatuses of transport must be fundamentally recast. As we have seen, within mobility studies approaches to reforming these systems range from the technofixes of the capitalist class and mainstream transportation policy makers, to the struggles for transport accessibility and traffic justice of liberals, to the decelerated, localized, autonomobility system of Cass and Manderscheid and radical sustainability advocates. Although the last of these approaches comes much closer in spirit to the general sensibility and philosophical perspective developed in the preceding chapters, it still fails to acknowledge the full effects of hypermobility on the more-than-human world or the myriad ways in which more-than-human beings contest and resist modern systems of mobility. Both of these issues carry profound consequences for how we envision the development of genuine altermobilities as well as the development of new forms of subjectivity and ways of life beyond the hypermobile present.

One of the few fields of research and practice that has sought to adopt something like this broad perspective on automobility—that is, a perspective that takes directly into consideration animal life/death and roadkill in the broad sense we have granted the term, as well as ecological relations and systems more generally—is the field of road ecology.[11] Road ecologists have as one of their primary tasks the development of mitigation strategies

[11] For an introduction to this field, see Coffin (2007), Davenport and Davenport (2006), Forman et al. (2003), and Ree et al. (2015). See also Gary Kroll's (2015) insightful article on how road ecology ushers in an alternative conception of the highway as permeable. I thank an anonymous reviewer for alerting me to Kroll's essay.

aimed at reducing roadkill and human-animal collisions. Some of these strategies involve relatively mundane practices such as posting signage to indicate the presence of wildlife near roads and placing fencing at strategic points along busy roadways. But road ecologists have also developed—and have even had some good fortune in instituting—more ambitious and more costly projects, such as large-scale wildlife crossings. Recently, near my own home, ground has been broken on one of the world's largest wildlife crossings in the United States, the Wallis Annenberg Wildlife Crossing, which is projected to cost some $90 million dollars and cross a ten-lane freeway in the Santa Monica Mountains. The area is populated by mountain lions and a host of other animals who migrate across these busy and dangerous roads, and this crossing should help substantially reduce the problem of roadkill in the area. Likewise, there is a statewide wildlife crossing bill currently under consideration in my home state of California, along with related bills and proposals that have been introduced at the federal level. Although such projects are primarily cast in terms of the safety and benefits that will accrue to human drivers (e.g., reductions in human-animal collisions will save on car repair costs, lower driver injuries, lower insurance premiums, and so on), the evidence we have at present suggests that wildlife crossings (coupled with fencing and similar measures) of this sort are generally effective at reducing roadkill and constitute one of the better tools that road ecologists can use to prevent wildlife-vehicle collisions (Huijser et al. 2009, Huijser et al. 2016).

As noted in Chap. 3, when mainstream animal rights groups address the problem of roadkill, they often tout such mitigation measures. Wildlife crossings in particular are popular with animal advocacy groups, inasmuch as they are seen as a win-win situation for motorists and animals: motorists can maintain their habitual patterns of driving, and animals can continue their normal migration patterns uninterrupted (or so the presence of these crossings might lead us to believe). The problem here, though, is precisely that which renders such solutions attractive to reformers—namely, that mitigation measures of this sort do little to challenge the status quo of hyperautomobility in any significant way. Mitigation measures can and do reduce roadkill but, in a certain sense, they can also further mystify and naturalize it—especially if such measures take for granted the current way of life associated with hyperautomobility (anthropocentric, capitalist, and so on) and work within its confines to ameliorate a small fraction of its worst effects.

Now, as someone who has been involved in pro-animal politics for many years and who is well aware of how challenging it is to achieve even minor pro-animal reforms, I sincerely applaud the technological reforms and innovations introduced by road ecologists and planners, and I am deeply grateful for the activists and lawyers who introduce the bills for mitigation projects and do the hard work of lobbying to get them passed. I have no intention (as one reviewer of this project cheekily suggested) of throwing the mitigation crowd "under the bus." That any attention is given to roadkill and that any efforts are made on behalf of the welfare of wildlife in relation to dominant transportation systems is remarkable, given the current configuration of the established order (Koelle 2012). That point noted, it must also be stressed that the task of philosophy is not simply to endorse and laud efforts at reform (any more than it should settle for throwing stones at passing buses, to extend the metaphor in a different direction). As with the approach taken by Cass and Manderscheid, a philosophical analysis of roadkill cannot limit itself to what is deemed pragmatic by the established order, any more than it can remain content with its limited social imaginary. Thus, we might consider following Cass and Manderscheid's lead and endeavor to extend our future imaginary a bit further, this time in a more deeply non-anthropocentric direction. If we grant the unsustainability of hyperautomobility; the economic inequality that plagues modern systems of mobility; the biodiversity loss, extinction crisis, and broader patterns of defaunation that attend hypermobile ways of life; the air, soil, and water pollution introduced into ecosystems and social systems by car use; and a whole host of related problems, then it is essential that we at least attempt to think and practice fundamentally non-hypermobile and non-anthropocentric ways of being-in and moving-through the world, even if they seem utopian or unrealistic at present.

If we pursue such a path in line with the inhumanist and non-anthropocentric perspectives developed in the preceding chapters, it becomes clear straight away that mitigation—while entirely necessary—is fundamentally inadequate for addressing the issues at hand. In the United States alone there are several million miles of roads and tens of thousands of miles of highways; even well-coordinated and well-funded efforts at providing fencing, crossings, and similar mitigation measures at key human-animal collision sites would reduce roadkill only by a small fraction. What is truly needed (but rarely acknowledged) in order to address roadkill and for both wildlife and all of planetary life to flourish (or at least to live and die in better conditions than many do at present) is to curb hyperautomobility and its concomitant way of life in a ***radical and sustained*** manner. This means not just better "management" of the existing system or mitigation of its harms, but fundamentally dismantling it: fewer people driving fewer miles in less complex vehicles (to reverse Freund and Martin's definition of hyperautomobility). This sort of deceleration (Vannini 2014), pausing (Sloterdijk 2020), and radical reversal of hypermobility would, to be sure, include the sorts of measures proposed by Cass and Manderscheid, but would go beyond them to investigate more basic issues such as: Which roads need to be temporarily and perhaps permanently closed so that vulnerable beings and systems can reclaim needed space and migration routes?[12] How can we best reserve and preserve the use of mobility infrastructure for those individuals and groups who most genuinely need it? Which sorts of practices and activism are most effective for dismantling hypermobile systems and ways of life? How to reconstitute more respectful relations with the more-than-human beings, systems, and other relations with whom we move? These are just a few of the kinds of questions concerning profanation that come to the fore when we think beyond system reform and when the human no longer constitutes the locus of meaning or the exclusive ground of sociality.

Ultimately, the goal of inhumanist, non-anthropocentric practices of profanation is to create a way of life where mobility requirements are drastically reduced and where it is possible for the bulk of people to live without cars and other features of hyperautomobile infrastructures, reserving those modes in limited forms for those who have genuine need of them. Such a goal might appear wildly implausible, but it should be noted that many people are ***already*** living in this manner (whether by choice or by necessity) and that the bulk of older people alive today lived in this way in their youth. In fact, a life of radically scaled-down and relatively slow mobility is all that human beings have known for millennia until just a handful of generations ago. The capitalist class and others invested in hyperautomobility do whatever they can to persuade us that the future is and must be ever-more hypermobile and that automobile use will continue to spread and take the form of ever-faster and ever-more electric and automated vehicles (Marx 2022). But there is nothing inevitable about this trajectory, and there is every reason—among the most important reasons being the welfare of both the human and more-than-human

[12] On the "hyperfragmentation" caused by road systems and the need to maintain and reinstitute roadless areas, see Trombulak and Frissell (2000).

worlds—fundamentally to contest that trajectory and not remain content just to manage it and mitigate its worst effects.[13]

Beyond these dismantling and decelerative measures, there is a concomitant need fundamentally to reorient our collective environmental sensibilities and mobilities in a more affirmative direction. By environment here I do not simply mean "nature" as the backdrop for human activities, but rather the environment as the dynamic and mobile *ēthos* (abode) we and other beings inhabit, the land in and through which life is sustained and value is co-constituted (Coulthard 2014). Environmentally minded altermobilities of this sort are not limited to critical efforts to avoid producing roadkill but are also aimed affirmatively at developing habits of mobility that are more consonant with and more respectful of more-than-human mobilities and ecologies (Ingold 2004; Larsen 2014). Such altermobilities go beyond standard forms of transport planning and urban re-design to consider the possibility of rewilding (or what I would prefer to call alter-wilding) particular regions such that differentially and relatively (im)mobile animal and more-than-human beings and systems have a chance at a sustained existence.[14] This attention to the variety of animobilities (Michael 2004) and other more-than-human mobilities intersects in productive ways with resurgent indigenous movements for the repatriation of land and for the ongoing development of indigenous mobilities that contest dominant modes of transport (Whyte et al. 2019). Indeed, at present, indigenous peoples and radical pro-animal and ecological activists

[13] The global COVID-19 pandemic, especially in its early stages when travel and other systems of mobility were radically curtailed, gave us a glimpse of the promises and challenges involved in such reductions and alterations. The changes brought on by the pandemic led to many clear benefits for both human and more-than-human beings and systems, but also increased injustices and problems for people who were vulnerable to rapid changes in mobility infrastructure. Reduced mobility, such as we saw in the initial stages of the pandemic, has the potential of course to turn into narrow localism and provincialism, but it might also offer the opportunity to rethink relation and sociality beyond hypermobile circuits. For more on these sorts of "compensatory" mobilities, see Nikolaeva et al. (2022). With regard to roadkill in particular, the pandemic clearly showed that early reductions in automobile use created a wider range of habitat for many animals; however, as driving began to return to pre-pandemic levels, the response by animals seems to have lagged that of human drivers, thereby lifting roadkill numbers to levels even higher than pre-pandemic levels (Abraham and Mumma 2021). What this phenomenon demonstrates is, again, the need for both radical and sustained curtailment of driving and related mobility systems across time and place, rather than the kinds of fluctuations that have occurred during the pandemic.

[14] See Lorimer (2015), Lorimer et al. (2015), and Bekoff (2014) for a deeper consideration of the political promises and limitations of rewilding.

are among the only groups who are keeping alive this vision of moving differently and in common with our more-than-human kin.[15]

Finally, reflecting on the event of roadkill recalls us to the vulnerability of embodied existence. Embodied vulnerability is, of course, something we share indistinctly with animals, though (as many of the roadkill artists examined in Chap. 3 propose) we often prefer to block this shared condition from consideration. But the stark consequences of dominant systems of mobility—the millions of animals who become roadkill annually, the tens of thousands of human beings who are killed in collisions every year, the millions who suffer from transport-related injuries and health issues—recurrently force this shared condition back into view. The field of disability studies has long sought to bring embodied vulnerability to the fore of theoretical and activist work in ways that are directly relevant to the foregoing considerations of mobility and animality. In the context of mobility studies, the point of such work by variously abled persons and organizations is not reducible to debates about mere accessibility to dominant systems of mobility but goes further to examine the ways in which individuals are problematically presumed to be able-bodied and largely immune to the many negative effects of automobility (Lubitow et al. 2017); this work also questions the assumed premise that all human beings are capable of and desire hypermobility and that the only way of life that is worthwhile is one in which individuals are in ever-more motion. This field of research and activism thus insists that worthwhile lives already exist outside of the circuits of hyperautomobility and that profanation of modes of mobility must amount to something more than expanding and reforming a fundamentally unjust and exclusionary set of (infra)structures and should include an affirmation of "differential mobilities."[16]

In the context of animal studies, disability theorists have made the case that an ethic starting from bodies that are understood to be vulnerable and always already differently abled provides the ground for a shared,

[15] While the respective visions and strategies of these groups are not identical, there is considerable overlap and significant promise for effective coalitions and alliances of various sorts. I examine such possibilities at more length in Calarco (2023).

[16] For more on the intersection of mobility studies and disability studies and related social movements, see Parent (2016), Sawchuk (2014) (from whom I borrow the term differential mobilities), and Vukov (2015). My remarks in this section have been especially influenced by Vukov's call for "building a multi-scalar and multi-modal transversal movement for mobility justice" that goes beyond the "polite norms of policy advocacy into broader and often less reformist-oriented social movements concerned with mobility" (Vukov 2015, 119).

alternative vision of ethical life between and among human beings and more-than-human kin (Jenkins et al. 2020; Wolfe 2010). What fundamentally ties us to our animal kin is not the existence of traces of the traits on which we pride ourselves (e.g., rationality, agency, intelligence) but the vulnerability and interdependence we share with and alongside them (Taylor 2011). To develop altermobilities from the perspective of this shared, profane space—a zone of indistinct life and death—requires us to delink ourselves from the individualist ideal of autonomy animating dominant modes of mobility and acknowledge our fundamental relationality and dependence on others (Sawchuk 2014), a condition shared with those who are both human and more-than-human and that unfolds in zones of life-death that are both terror-filled and majestic. It also requires us to acknowledge that the systems of mobility we have developed structurally occlude such dependence and interdependence from view, encouraging us to overlook the harms those systems cause and the ways in which bodies of various sorts are fundamentally incompatible with its ongoing function and expansion. It is in these senses that the event of roadkill is essential and crucial for catching sight of the dark underside of modern mobility. Resisting and contesting this underside does not require of us permanently to cease moving. Nor does it ask us entirely to eliminate the killing of animals and other beings (ourselves included). What it does require of us is to decelerate, to pause, even to come temporarily to a standstill in order to reflect on the ways in which the countless lives and deaths sacrificed to sustain an anthropocentric, hyperautomobile way of life might become something more and something other than mere collateral for a select class of sacred beings.

Coda

During the time I was completing the final edits for this work, I was invited to attend a release event at a local raptor recovery and rehabilitation center. The center is located on the edge of the city near a wilderness area and is accessible only by car via a winding dirt road. My family and I, along with about thirty other people, drove out to the center to watch the rehabilitated birds be released. The center encourages donors to help with the releases as a means of putting people in closer contact with the birds (the assumption being that donors will form deeper emotional bonds with the animals through personal proximity). Before the releases took place, the director of the center and the head veterinarian spoke to the assembled crowd about the work they do, the birds themselves, and the various threats to the birds' wellbeing. The talk was informative and covered such threats to raptors as the habitat fragmentation associated with urban sprawl, poisoned prey animals that the birds might consume, large glass buildings they might unwittingly strike, and the myriad effects on birds of the changing climate. The speakers encouraged the audience to think about what they could do to make changes in these areas and inspire others to do the same.

As the birds were displayed and carried out one by one to be released, the audience was clearly in awe. Among the rehabilitated birds were red-tailed hawks, Cooper's hawks, great-horned owls, and a falcon, each one of whom was distinct, dignified, and beautiful in its own way. As each bird was shown to the crowd, the director and veterinarian said a word or two about that particular bird and how it had ended up at the center. All seven birds released that day had, it turns out, been struck and injured by cars, which led to their being taken to the raptor center. As the speakers explained, when raptors spot their prey, they focus intently on it and fly in an efficient and direct path to strike it, paying little attention to anything else in their surroundings, including fast-moving cars.

Once the releases were completed and the birds flew back into the hills and trees on the edges of the canyon, the audience dispersed, returned to their cars, and drove back along the dirt road to the surrounding freeways and then back to their homes. There never was any mention, either from the speakers or from the generous and supportive members of the audience, of the need to change our driving habits to prevent injuring birds, or of the fact that our means for attending the raptor release event (namely, driving our cars) were fundamentally at odds with the flourishing of the animals everyone sought to promote. Perhaps such contradictions are too difficult for us to acknowledge; at some point, though, we will need to address these incongruities that have haunted us for so long now.

References

Abraham, Joel O., and Matthew A. Mumma. 2021. Elevated Wildlife-Vehicle Collision Rates During the COVID-19 Pandemic. *Scientific Reports* 11: 20391. https://doi.org/10.1038/s41598-021-99233-9.

Agamben, Giorgio. 2009. What Is an Apparatus? In *What Is an Apparatus?*, trans. David Kishik and Stefan Pedatella, 1–24. Stanford, CA: Stanford University Press.

Banister, David. 2005. *Unsustainable Transport: City Transport in the New Century*. New York: Routledge.
———. 2008. The Sustainable Mobility Paradigm. *Transport Policy* 15 (2): 73–80.
Bekoff, Marc. 2014. *Rewilding Our Hearts: Building Pathways of Compassion and Coexistence*. Novato, CA: New World Library.
Bissell, David. 2007. Animating Suspension: Waiting for Mobilities. *Mobilities* 2 (2): 277–298.
Calarco, Matthew. 2023. *The Three Ethologies*. Chicago: University of Chicago Press.
Cass, Noel, and Katharina Manderscheid. 2019. The Autonomobility System: Mobility Justice and Freedom Under Sustainability. In *Mobilities, Mobility Justice and Social Justice*, ed. Nancy Cook and David Butz, 101–115. New York: Routledge.
Chrulew, Matthew, and Dinesh Joseph Wadiwel, eds. 2017. *Foucault and Animals*. Leiden, The Netherlands: Brill.
Coffin, Alisa W. 2007. From Roadkill to Road Ecology: A Review of the Ecological Effects of Roads. *Journal of Transport Geography* 15 (5): 396–406.
Coulthard, Glen Sean. 2014. *Red Skin, White Masks: Rejecting the Colonial Politics of Recognition*. Minneapolis: University of Minnesota Press.
Cresswell, Tim. 2010. Towards a Politics of Mobility. *Environment and Planning D: Society and Space* 28 (1): 17–31.
Davenport, John, and Julia L. Davenport, eds. 2006. *The Ecology of Transportation: Managing Mobility for the Environment*. Dordrecht, The Netherlands: Springer.
Deleuze, Gilles. 1995. Postscript on Control Societies. In *Negotiations, 1972–1990*, 177–182. New York: Columbia University Press.
Dennis, Kingsley, and John Urry. 2009. *After the Car*. Malden, MA: Polity.
Estes, Nick. 2019. *Our History Is the Future: Standing Rock versus the Dakota Access Pipeline, and the Long Tradition of Indigenous Resistance*. New York: Verso.
Forman, Richard T.T., Daniel Sperling, John A. Bissonette, et al., eds. 2003. *Road Ecology: Science and Solutions*. Washington, DC: Island Press.
Foucault, Michel. 1980. The Confession of the Flesh. In *Power/Knowledge: Selected Interviews and Other Writings 1972–77*, ed. Colin Gordon, 194–228. New York: Pantheon.
Furness, Zack. 2010. *One Less Car: Bicycling and the Politics of Automobility*. Philadelphia: Temple University Press.
Heidegger, Martin. 1962. *Being and Time*. Translated by John Macquarrie and Edward Robinson. New York: Harper & Row.
Huijser, Marcel P., John W. Duffield, Anthony P. Clevenger, et al. 2009. Cost–Benefit Analyses of Mitigation Measures Aimed at Reducing Collisions with Large Ungulates in the United States and Canada: A Decision Support Tool. *Ecology and Society* 14 (2) http://www.jstor.org/stable/26268301.
Huijser, Marcel P., Elizabeth R. Fairbank, Whisper Camel-Means, et al. 2016. Effectiveness of Short Sections of Wildlife Fencing and Crossing Structures

along Highways in Reducing Wildlife–Vehicle Collisions and Providing Safe Crossing Opportunities for Large Mammals. *Biological Conservation* 197: 61–68.

Ingold, Tim. 2004. Culture on the Ground: The World Perceived Through the Feet. *Journal of Material Culture* 9 (3): 315–340.

Jenkins, Stephanie, Kelly Struthers Montford, and Chloë Taylor, eds. 2020. *Disability and Animality: Crip Perspectives in Critical Animal Studies.* New York: Routledge.

Koelle, Alexandra. 2012. Intimate Bureaucracies: Roadkill, Policy, and Fieldwork on the Shoulder. *Hypatia* 27 (3): 651–669.

Kroll, Gary. 2015. An Environmental History of Roadkill: Road Ecology and the Making of the Permeable Highway. *Environmental History* 20 (1): 4–28.

Larsen, Jonas. 2014. (Auto)Ethnography and Cycling. *International Journal of Social Research Methodology* 17 (1): 59–71.

Lorimer, Jamie. 2015. *Wildlife in the Anthropocene: Conservation after Nature.* Minneapolis: University of Minnesota Press.

Lorimer, Jamie, Chris Sandom, Paul Jepson, et al. 2015. Rewilding: Science, Practice, and Politics. *Annual Review of Environment and Resources* 40 (1): 39–62.

Lubitow, Amy, Jennifer Rainer, and Sasha Bassett. 2017. Exclusion and Vulnerability on Public Transit: Experiences of Transit Dependent Riders in Portland, Oregon. *Mobilities* 12 (6): 924–937.

Manderscheid, Katharina. 2014. The Movement Problem, the Car, and Future Mobility Regimes: Automobility as Dispositif and Mode of Regulation. *Mobilities* 9 (4): 604–626.

Martens, Karel. 2016. *Transport Justice: Designing Fair Transportation Systems.* London: Routledge.

Marx, Paris. 2022. *Road to Nowhere: What Silicon Valley Gets Wrong about the Future of Transportation.* New York: Verso.

Michael, Mike. 2004. Roadkill: Between Humans, Nonhuman Animals, and Technologies. *Society & Animals* 12 (4): 277–298.

Montegary, Liz, and Melissa Autumn White, eds. 2015. *Mobile Desires: The Politics and Erotics of Mobility Justice.* New York: Palgrave Macmillan.

Newman, Peter, and Jeffrey R. Kenworthy. 2015. *The End of Automobile Dependence: How Cities Are Moving Beyond Car-Based Planning.* Washington, DC: Island Press.

Nikolaeva, Anna, Ying-Tzu Lin, Samuel Nello-Deakin et al. 2022. Living Without Commuting: Experiences of a Less Mobile Life Under COVID-19. *Mobilities.* https://doi.org/10.1080/17450101.2022.2072231.

Parent, Laurence. 2016. The Wheeling Interview: Mobile Methods and Disability. *Mobilities* 11 (4): 521–532.

van der Ree, Rodney, Daniel J. Smith, and Clara Grilo, eds. 2015. *Handbook of Road Ecology*. Hoboken, NJ: Wiley-Blackwell.

Sawchuk, Kim. 2014. Impaired. In *The Routledge Handbook of Mobilities*, ed. Peter Adey, Mimi Sheller, David Bissell, Kevin Hannam, and Peter Merriman, 409–420. New York: Routledge.

Scott, Nicholas A. 2020. *Assembling Moral Mobilities: Cycling, Cities, and the Common Good*. Lincoln: University of Nebraska Press.

Sheller, Mimi. 2018. *Mobility Justice: The Politics of Movement in an Age of Extremes*. London: Verso.

Sloterdijk, Peter. 2020. *Infinite Mobilization: Towards a Critique of Political Kinetics*. Translated by Sandra Berjan. Medford, MA: Polity.

Stehlin, John G. 2019. *Cyclescapes of the Unequal City: Bicycle Infrastructure and Uneven Development*. Minneapolis: University of Minnesota Press.

Stiegler, Bernard. 2010. *Taking Care of Youth and the Generations*. Translated by Stephen Barker. Stanford, CA: Stanford University Press.

Swart, Sandra. 2015. Reviving Roadkill? Animals in the New Mobilities Studies. *Transfers* 5 (2): 81–101.

Taylor, Sunaura. 2011. Beasts of Burden: Disability Studies and Animal Rights. *Qui Parle: Critical Humanities and Social Sciences* 19 (2): 191–222.

Tolley, Rodney. 2010. *Sustainable Transport: Planning for Walking and Cycling in Urban Environments*. Boca Raton, FL: Woodhead Publishing Limited.

Trombulak, Stephen, and Christopher Frissell. 2000. Review of Ecological Effects of Roads on Terrestrial and Aquatic Communities. *Conservation Biology* 14 (1): 18–30.

Vannini, Phillip. 2014. Slowness and Deceleration. In *The Routledge Handbook of Mobilities*, ed. Peter Adey, Mimi Sheller, David Bissell, Kevin Hannam, and Peter Merriman, 116–124. New York: Routledge.

———, ed. 2016. *The Cultures of Alternative Mobilities: Routes Less Travelled*. New York: Routledge.

Vukov, Tamara. 2015. Strange Moves: Speculations and Propositions on Mobility Justice. In *Mobile Desires: The Politics and Erotics of Mobility Justice*, ed. Liz Montegary and Melissa Autumn White, 108–121. New York: Palgrave Macmillan.

Whyte, Kyle, Jared L. Talley, and Julia D. Gibson. 2019. Indigenous Mobility Traditions, Colonialism, and the Anthropocene. *Mobilities* 14 (3): 319–335.

Wolfe, Cary. 2010. *What Is Posthumanism?* Minneapolis: University of Minnesota Press.

Index[1]

[1] Note: Page numbers followed by 'n' refer to notes.

M. Calarco, *Reflections on Roadkill between Mobility Studies and Animal Studies*, https://doi.org/10.1007/978-3-031-30578-8

GPSR Compliance
The European Union's (EU) General Product Safety Regulation (GPSR) is a set of rules that requires consumer products to be safe and our obligations to ensure this.

If you have any concerns about our products, you can contact us on

ProductSafety@springernature.com

In case Publisher is established outside the EU, the EU authorized representative is:

Springer Nature Customer Service Center GmbH
Europaplatz 3
69115 Heidelberg, Germany

www.ingramcontent.com/pod-product-compliance
Ingram Content Group UK Ltd.
Pitfield, Milton Keynes, MK11 3LW, UK
UKHW020258051026
15470UKWH00021B/1397
* 9 7 8 3 0 3 1 3 0 5 7 7 1 *